The Design, Construction, Testing, and Flight of a

Bipropellant Liquid Fuel Rocket

The Design, Construction, Testing, and Flight of a

Bipropellant Liquid Fuel Rocket

by

George D. Morgan

Published by Pony-X Press
Los Angeles, New York, Dallas
Copies of this report are available through Amazon.com.

For more information, contact the author through his web site: www.georgedmorgan.com

Books by the author (available through Amazon, Barnes & Noble, and your local bookseller):

- Rocket Girl: The Story of Mary Sherman Morgan, America's First Female Rocket Scientist

- Rocket Age: The Race to the Moon and What it Took to Get There

- Play Dead

- Moon Hunter

This report may not be duplicated in whole or in part without written permission from the author. Requests for such should be directed to the author through his website: www.georgedmorgan.com, or by email: wgawriter@aol.com.

Permission need not be obtained, however, for the physical reconstruction of the rocket described in this report. Such permission is hereby granted, without royalty, and the engineering drawings have been provided for this purpose.

All technology described in this report is in the public domain. Therefore, the security classification of the rocket described herein is **UNCLASSIFIED.**

TABLE OF CONTENTS

THIRST FOR LIQUIDS

It was February 1980. Dan Ruttle, an aerospace engineer and close friend, and I were having another one of our many discussions about liquid fuel rockets. Why, we wondered, don't more experimental rocketeers build liquid fuel rockets? Our musings were, of course, ironic, given that neither of us had ever built one.

In the late 1940's, members of the Pacific Rocket Society were the first private experimenters to dabble in hybrid rockets. We believe they successfully flew at least one, but detailed records of their projects are non- existent or unavailable.

In the 1950s Dave Elliot and Lee Rosenthal flew their hydrogen peroxide rocket near Rosamond, California. Around the same period of time a Canadian organization launched a small acid/alcohol rocket called Star I, then faded into obscurity. We know of no liquid projects during the 1960s. Then in 1976 David Crisalli[1], a midshipman at Annapolis, built and flew an ambitious 8-inch diameter, 18-foot-long Kerosene and Oxygen rocket at the White Sands Proving Grounds as a class project for the Naval Academy. In 1982 his project was duplicated, albeit on a slightly smaller scale, by another Naval Academy student, Ben Loyola. Finally, the School of Engineering at Redlands University has flown at least one hybrid rocket that we know of.

So as of the original publication date of this report—December 1987—that makes six privately built rockets that utilized liquid propellants in a span of thirty years.[2]

However, as a direct result of the publication of the first edition of this report in 1987, there have been dozens of liquid propellant rockets flown in the U.S. since then. In fact, there have been so many projects we gave up keeping track of them. One of the most notable projects was by an Australian undergraduate engineering group from Monash University known as AUSROC (for Australian Rocketry). On February 9 1989 they built and flew a modified version of the rocket described in this report.[3]

Despite the surge in activity, however, home-built liquid fuel rockets have a long way to go before they are as popular as video games. We decided to write down all the reasons for this glaring lack of activity and came up with a list of the eight main challenges to successful liquid propellant rocket construction:

---

[1] David Crisalli and I were in the same class at Chaminade Preparatory in West Hills, California (Class of 1971). At the time, Chaminade had one of the strongest and most advanced high school experimental rocket clubs in the world.

[2] This, of course, applies only to the United States. As the internet had not yet been invented, we did not have easy access to projects outside the country.

[3] Information of AUSROC at one time could be found with a Google search. Alas, no longer.

1. Plumbing is for Plumbers. Flowing liquids require plumbing: pipes, fittings, valves, seals, tanks, pressure systems, etc. Plumbing creates design complications sufficient enough to scare off most experimental rocketeers all by itself.

2. Fabrication Nightmares. Machining, metal bending, precision welding, airframe construction, the drilling of minute injector holes and a myriad of other fabrication headaches keep most rocketeers cozy and warm at home with their solids

3. Embarrassing Mass Ratios. Unless you're building an Atlas or Saturn V, getting a respectable mass ratio with liquids is a major challenge. Experimental rocketeers, who work on a much smaller scale, logically conclude, "Why bother?"

4. The Candle Won't Light. Ignition is a bigger problem with liquids. The professional aerospace record is replete with ignition headaches.

5. You Can't Get LOX Tanks at Walmart. The higher cost of key components (including most propellants) and the struggle to find them on the open market can be the biggest roadblock of all. Who wants to pay $300 for a check valve? (especially when you need *two* of them!)

6. Who has the Time? A solid propellant rocket can be built in one day. The project you are about to read took seven years of on-again, off-again work. That level of patience and persistence is hard to find in *any* hobby.

7. Lack of Good Instructional Data. The How-To Book of Liquid Propellant Rockets has never been written (as far as we are aware). Where does one go to get information on how to design and construct these things? George Sutton's book is more of a college textbook, not a practical hands-on guide for the experimental builder. Much of the data necessary for liquid propellant rockets is either unpublished or unavailable.

8. Greater Complexity = Lower Reliability. After all the time invested, after all the expense, after all the years of work, you finally push the button, only to have it explode on the launch rack. You ask yourself, "Am I having fun, yet?" Your friends wisely advise you to stick to stamp collecting.

No wonder almost nobody wants to build these things. And yet, as many experimental rocket enthusiasts will tell you, the liquid propellant rocket looms ever present on the distant horizon, taunting us with its challenge. Most mountain climbers who spend a few years bagging peaks in the High Sierras or Rockies will eventually start looking south to Patagonia, or east to the

Himalayas—searching for a greater challenge for their acquired skills. So it is with experimental rocket builders.

For those who thirst after liquids like we do, this report will offer a respite on a dry spell that has lasted too long. For the others who only shake their heads and wonder why we would spend so much time and money on such an innocuous and unprofitable venture, I can offer no better explanation or simple rationale than the quote so often misattributed to Sir Edmund Hillary: "Because it's there."

"George, we're going to do it," declared Dan. And having thus spoken, the dream of building a liquid rocket became a goal.

We had been going over our list of "whys", i.e. why people don't build liquids, and had settled on several basic concepts of the design to overcome most of the inherent problems.

First, we would overcome the plumbing problem in a very simple way: we would eliminate the plumbing! Our rocket would have concentric tanks (one nestled inside of the other) which would bolt directly to the injector. Both tanks would have a pre-pressurized full-flow start, eliminating the need for a separate on-board pressure vessel.[4]

Second, we would resolve the fabrication problem by keeping the rocket small in size. The lack of plumbing would have the side benefit of aiding in the fabrication, and a small injector would mean fewer precision holes to drill. The engine would use an ablative liner. The motto for the project would be: **Keep It Simple.**

Third, we would solve the mass ratio problem by ignoring it. No rocket has everything, and everything has a trade-off. We would build this liquid rocket solely for the education and experience, not for the performance.

Fourth, ignition would be made more reliable by using hypergolic propellants.

Fifth, all components would be either off-the-shelf hardware or simple enough that we would be able to fabricate them ourselves.

Sixth, by making a firm commitment to finish the project no matter how long it took, we determined to defeat the time enemy.

Seventh, we would resolve the technology problem by pooling all of our resources. Though Dan Ruttle did the overall engineering design work, many individuals supplied solutions to technology problems that cropped up along the way.

Eighth, the reliability problem would be resolved, we felt, by strictly adhering to the aforementioned project motto: Keep It Simple. Complexity would be avoided like the plague.

---

[4] The downside to this design choice is that the propellant tanks could only be 2/3 full as ullage would be needed as space for the pressurized gaseous nitrogen.

Against my strongly worded advice, Dan decided that the Keep It Simple rule would preclude the inclusion of a parachute recovery system. To him it was just one more thing to design, build, and test, and therefore a needless luxury. It was a decision that would come back to haunt us.

So. Keep It Simple. Fine. But in the back of our minds we all wondered the same thing. It was an obvious question—one that begged to be asked. Yet one that nobody chose to vocalize at the time. What on Earth does a "simple" injector look like?

ROCKET FUEL: YOU CAN'T LEAVE EARTH WITHOUT IT

Before even the first nut and bolt could be designed, and certainly before the injector could be, we needed to choose which propellants we would use. Here is how we selected the propellant combination.

Currently there are three basic types of liquid-fuel rocket designs: monopropellant, bipropellant, and hybrid.

A monopropellant rocket, as the prefix implies, utilizes a single propellant, such as hydrogen peroxide, to produce thrust. In an H_2O_2 rocket, for example, the propellant is decomposed rapidly via a pyrotechnic device, a heating element, or exposure to chemical impurities like a catalyst bed. This decomposition produces steam and oxygen gas and the resultant expansion gives the rocket its thrust. While monopropellants have the advantage of simplifying the rocket's design, they carry the serious disadvantage of having to use unstable chemicals. They also tend to be relatively low in performance.

In a bipropellant rocket, a fuel and an oxidizer are carried aboard in separate tanks. A plumbing system feeds the propellants into an injector and combustion chamber where they are mixed and burned. Bipropellants have the highest theoretical performance of the three designs, however in reality the greater complexity and weight required for the rocket's airframe can result in serious trade-offs.

In a hybrid system, a liquid propellant, usually an oxidizer such as hydrogen peroxide or liquid oxygen, is sprayed into a combustion chamber lined with a solid propellant, usually a fuel such as paraffin or rubber. Though hybrids are better performers than monopropellants and are also simple in design, they are generally lower in performance than bi-propellants due to the inefficiency in mixing the fuel and oxidizer. Since we had decided on a bi-propellant design for our project, we then had to choose which of the many fuels and oxidizers available we would use. This choice was made simpler by our requirement for hypergols, since only a narrow range of rocket propellants fit that category.

First, the oxidizers.

OXIDIZERS WITH HYPERGOLIC MATES	ADVANTAGES/DISADVANTAGES
Fluorine	Very high performance. Special handling required due to extreme reactivity. High cost. Cryogenic. Very toxic.
Chlorine Trifluoride	High performance. Special handling required due to extreme reactivity. Very toxic.
Chlorine Pentafluoride	High performance. Special handling required. High cost. Very toxic.
Nitrogen Tetroxide	High performance. High cost and availability strictly controlled. Very toxic.
Nitric Acid	Medium performance. Low vapor pressure. Low availability in the required strengths. Medium cost.
Red Fuming Nitric Acid[5]	Medium performance. Low availability in required strengths. High cost and extreme toxicity. Toxic fumes.
Hydrogen Peroxide	Medium performance. Low vapor pressure. Explosion hazard and handling problems. Very low availability in high strengths.

The number of fuels with hypergolic mates is many times greater than the oxidizers. Because of this it is common in hypergolic systems to choose the oxidizer first and the fuel second. This is exactly the procedure we followed as well.

Our choice was made by a simple process of elimination. Due to cost and handling problems we immediately eliminated fluorine, the two chlorine compounds and nitrogen tetroxide. Hydrogen Peroxide's explosion hazard was enough to eliminate it even if it were available in the U.S. (which it was not in 1987). Toxicity eliminated red fuming nitric acid.

---

[5] Two weeks before the 2020 edition of this report was published, Irving Kanarek—former attorney for Charles Manson—passed away at age 100. Little known fact: long before he was an attorney, Kanarek was a chemical engineer for North American Aviation and invented red fuming nitric acid for the US Army Nike program.

That left nitric acid. Though quite toxic in its own right, its vapors were not as deadly as some of the others. In small amounts we would be able to handle it. Plus, it had an added advantage that was important to us; we could work with it at ambient temperatures and pressures. This was necessary if we were to Keep It Simple.

And so nitric acid became our oxidizer. This oxidizer is sometimes referred to as white fuming nitric in order to avoid any confusion with red fuming nitric. However, because the term "white fuming" has largely fallen into disuse, we will avoid its mention in this report.

The fuels.

FUELS HYPERGOLIC WITH NITRIC ACID	ADVANTAGES/DISADVANTAGES
Hydrazine	High performance. High toxicity, low availability, high cost.
Furfuryl Alcohol	Medium performance. Low toxicity. Medium cost. Ease in handling.
Diethel Triamine, Methel Amine, et al	Good performance. Low availability. Very toxic.
Aniline	Medium performance. Low cost and good availability. Very toxic. Low vapor pressure.

Its high cost, toxicity, and low availability eliminated hydrazine and the amine compounds. We seriously considered aniline, which is the traditional hypergolic companion of nitric acid. The WAC Corporal and other early missiles enjoyed significant success with that combination. However, there was one other important factor we used in making our decision: ignition delay. Nitric acid and furfuryl alcohol have an ignition delay of twenty milliseconds, while nitric acid and aniline's ignition delay is one hundred milliseconds. Since this rocket would have a pre-pressurized full-flow start, the faster ignition characteristic was desirable.

For us, that tipped the scales; furfuryl alcohol would be our fuel.

OFF THE DRAWING BOARD

With the selection of the propellants completed, Dan now commenced setting the design parameters. He settled on the following:

Thrust	300 lbs.
Burn time	6 seconds
Tank pressure	400 psi
Maximum diameter	4 inches (10.16 cm.)
Total length	77 inches (197.58 cm.)

For the next six months I heard nothing out of Dan. It was as if he had locked himself away in a cave or something. All during this period, as I learned later, he was spending every spare moment designing and performing the engineering work on the rocket. Each part was carefully engineered and slowly, bit by bit, the rocket began to take form on paper. By the end of that six-month period, Dan had a full set of drawings and was ready to start "bending metal."

Dave Griffith, a fellow rocket enthusiast and owner of a machine shop east of Los Angeles offered his services and equipment for the project. Because of the distances involved between my home and his shop, I was not involved in this part of the project. Two or three times a month Dan would visit Dave's machine shop to work on the propellant tanks, bulkheads, adaptor tube, pressure fittings, pyro-valves and the injector.

Over the next year a great amount of work was accomplished in a relatively short time. It began to look like we would be able to launch in early 1982.

Then something happened that would delay the project almost five years. The injector, which was the most precision component of the rocket was 95% completed. All that was left to do was drill eight fuel holes and eight oxidizer holes and the most difficult part of the entire rocket would be complete. While machining the injector holes, Dan broke a drill bit and was unable to pull the bit out of the hole. On the next hole a second bit broke off. Then a third. So frustrated and angry was he at this failure that he took the tanks and all the components he had been working on and mothballed them. In essence, Dan shelved the project and returned to his normal daily life.

The rocket, in its many pieces and unfinished parts, sat around Dan's apartment for a number of months. Then one day he called me and asked if I wanted to hold onto them and store them in my garage. Wanting urgently to protect the project and its components I agreed, and so in February of 1982 the transfer was made. I set aside a small corner of my garage for the tankage and a box of parts.

And there the project sat collecting dust. For years.

RENEWED ENTHUSIASM

In December of 1985 I made what would turn out to be an historic decision. The Pacific Rocket Society—a non-profit corporation founded in the 1940's by George James—had fallen on hard times and had been defunct for over ten years. Its former members had scattered to the four winds. It was at this time that I decided to reactivate the PRS and reinstate it as a legal non-profit corporate entity. After numerous letters and phone calls to the appropriate state legal agencies and departments, the Pacific Rocket Society, Inc. was reinstated. Before long I received a removal-from-suspension notice from the Franchise Tax Board and we were back in business[6]

With the advantage of a corporate identity and non-profit legal status came renewed enthusiasm for the liquid rocket project. At the beginning of 1986 I called Dan at work and made an observation. "If you're going to be in the rocket business, you need a lathe."

So, we bought one.

A few weeks later we had a 12-inch Frejoth sitting in the garage of my parents' home at 8444 Melba in West Hills, CA. That was March of '86. By May a flurry of activity was in progress. Most of this activity, however, was not related to the liquid rocket. Much of it was on solid rocket projects. But by the summer of '86 I had made the decision to make some real progress. One Saturday morning I arrived at the house with a singular goal in mind: To finish the engine. It wasn't going to be easy. The night before I had called Dan and asked him for a copy of the engineering drawings of the liquid rocket engine. His response took me by surprise.

"I don't have them," he had said.

"You don't have them? Why not?"

"They were stolen quite a while ago."

"Stolen! By who? How? Why?"

He explained how one day, while leaving his apartment, he had lain the complete (and only) set of drawings outside on the air conditioning unit of his apartment for a moment while he went back inside to retrieve something. When he came out, the drawings were gone. But the engine had to be built. And so, with a basic idea in my head of what it was supposed to look like, I went to work. The project, sans drawings, was back in gear.

---

[6] The Reaction Research Society—another Southern California rocket club—would later attempt to block this reinstatement out of jealousy, envy, and spite. Their numerous attempts to undermine us proved unsuccessful. As of 2025, the Pacific Rocket Society remains in full force.

THE ENGINE

"Only someone who has made 65,000 errors is
qualified to build a rocket."

—Wernher von Braun

During the years of inactivity on the rocket, the engine lay in pieces and raw materials in a cardboard box. Along with the main body, I took it with me every time we moved, which was quite often. Many personal possessions would find themselves in the garbage before each one of these moves, but the tanks and the engine parts were inviolate; they were always the first items to be packed.

The engine design consisted of a 10.5-inch steel tube, 1/16-inch wall thickness and 3-inch O.D. Inside of this tube would be epoxied a linen-phenolic ablative liner. This liner was a tube itself, 2-3/4 inches O.D. and .33-inch wall thickness. It was 8 inches long, this to allow a recess on each end, one for the injector, one for the nozzle. The nozzle would be steel with a graphite insert and the injector would be machined from a single block of 6061-T6 aluminum. Both would be bolted to the steel tube. Finally, fin tabs would be welded to the outside of the completed engine.

Measurements were taken as to the amount of recess needed on both ends of the tube. Then the ablative liner, which was about two inches too long, was cut the approximate length, leaving a little extra to be machined off later. The liner was then epoxied into the steel combustion chamber tube without incident.

After hardening of the epoxy bond, the entire assembly was chucked up in the new lathe and the excess liner was cleaned up from each end. The holes for the nozzle were then drilled and tapped. A preliminary fit-up revealed a gap between the injector face and the liner edge, but of negligible size.

The injector at this point was in a state of 95% completion, machined by Dan at Dave Griffith's shop over five years previous. Though the basic design eventually proved more than satisfactory, the decision to leave the drilling of the injector holes to last was flawed. Lesson learned: do the hardest job first.

For months we had been talking about how to save the damaged injector. One day I decided to let a professional handle it and I dropped it off at a nearby machine shop where I was living in Santa Paula. I laid it on the owner's desk and told him what I needed. He said he'd have it done in a few days. Weeks later it was still sitting on his desk, so I took it back.

Years of inactivity would follow the damaging of the injector, but on that day in 1986 our enthusiasm for finishing the project was high enough to overcome any obstacle.

Dan had machined the nozzle to the point where its outside diameter was correct, the cavity for the graphite insert was completed, and the exit angle started. So next the exterior angle was machined to its proper design angle and length. Then the plug of graphite was epoxied into the cavity, after shaving off several thousandths for a better fit.

Upon hardening of this second epoxy bond the exit end was chucked up in the lathe and a one-inch hole drilled through the graphite for the throat, followed by the machining of the entrance angle. Except for the repair of the injector, we now had a liquid fuel rocket engine. Eventually the combustion chamber was taken to Sheetcraft in Santa Paula where Bill Mensing[7] welded on the fin tabs for the aluminum fins.

Upon seeing the welded fin tabs with the fins bolted on, Dan remarked, "That's the best fin attachment I've ever seen."

---

[7] Bill was my father-in-law and is famous for having invented the under-belly water-bombing tanks for fire-fighting helicopters. Sadly, Bill passed away in 2012.

THE BLOWDOWN SYSTEM

Commercial liquid fuel rockets need a method of forcing the propellants at high velocity into the rocket engine. There are two classic design choices for this:

- Pressure Vessel. A third tank, containing an inert gas such as nitrogen or helium, is mounted above the propellant tanks. Moments prior to ignition a valve is opened, and the pressure vessel's gaseous contents are fed into the propellant tanks, which force the propellants into the engine. This system is usually used in smaller rockets such as the WAC Corporal.

- Turbo Pumps. In this design the propellants are pumped into the engine rather than pressure-fed. Turbo pumps are usually used in large rocket systems like the Atlas, Saturn V, and the Space Shuttle.

Both of these systems, we knew, had the same big disadvantage: they violated the Keep It Simple rule. Dan decided to design the rocket with a third, rarely used, pressure system rocket engineers refer to as the "blowdown" method. In a blowdown system the propellant tanks are pre-pressurized before launch by an external tank that remains on the ground after takeoff. At launch, valves at the engine open up and allow the already-pressurized propellants to flow into the combustion chamber. There are lots of advantages to such a system, but there is one important disadvantage: there has to be room left in the tanks for the pressurized gas. As a result, you cannot fill the propellant tanks all the way. In most cases (as with ours) 1/3 of the volume of the tanks had to be left empty as ullage to make room for the pressurization gas. Blowdown systems decrease a rocket's efficiency and performance, but they sure do make them a great deal easier to build.

It was Dan on the phone. As we exchanged pleasantries, I detected a note of sadness in his voice. Or was it depression?

"What's the problem old buddy?" I queried.

"Well, I found a supplier for the nitric acid."

"That's great! But why so glum? Price too high?"

"No, the price is fine."

"Any problem with selling it to us?"

"No. They're eager to sell it to us."

"Is shipping going to be a problem?"

"No, shipment is no problem."

"Then what's the problem?"

"It's the furfuryl alcohol."

I was getting impatient. "Is the price out of line?"

"No, the price is in line."

"Is there a shipping problem?"

"No, not at all. It can be shipped by UPS."

"Then what's the problem?!" I was getting frustrated—Dan was unhappy about something. I could hear it in his voice.

"Well..." There was a long pause. "It's Aldrich Chemical. The supplier. They refuse to sell it to us."

"On what grounds?" I asked, already suspecting the answer.

"On the grounds that we're going to use it in a rocket, and they don't want to be responsible if something goes wrong."

One of the cardinal rules of experimental rocketry is never tell a supplier what you're using their product for. If you do, they'll think you're a nut, a terrorist, or both. Dan, being so honest, always had trouble with this rule.

"No problem, Dan. I'll take care of it," I promised, getting the phone number of Aldrich Chemical from him. "Oh, and Dan," I added. "What else is furfuryl alcohol used for besides a liquid fuel rocket propellant?"

He thought for a moment, then replied, "The only thing I've ever heard it being used for is as a propellant."

Then he came up with what would prove to be a brilliant idea.

"If they ask you what you're using it for," he added, "tell them it's for acid catalyzed reactions."

"Acid catalyzed reactions." Utterly innocuous. Perfect.

Getting around arbitrary and bureaucratic rules had become old hat for me. In college once, for example, I decided to throw a party for some friends. The problem: how to get beer for fifty people when I was far from being legal drinking age. The solution: I called a nearby liquor wholesaler and identifying myself as a local businessman (that was true; my father and I owned a small manufacturing business together) I put in an order for two kegs of beer and told the order desk that I would "send one of my employees down to pick it up." Getting the total price from them, I then made out a company check for the amount, typed up an official purchase order, donned a pair of employee overalls, grabbed an official looking clipboard, and hopped in one of the company trucks. A few minutes later I was backing up to the distributor's loading dock.

"I'm here to pick up two kegs of beer for my boss," I announced in my best just-doing-my-job impression.

"Over there," someone replied, pointing. "Sign here."

As I signed the receipt form, two of their workers loaded the kegs onto my truck. Handing their supervisor the check, I thanked them for the help, then drove off, straining to keep a straight face until I was out of the parking lot.

Now I was faced with a similar problem. Ironically, like the party many years previous, the success of the entire project hinged on using subterfuge to purchase a form of alcohol.

"Well, George, let's see if you still have it in you."

I hung up with Dan and dialed Aldrich Chemical. A receptionist answered and I asked for the order desk. Soon a young man came on the other end of the line. I spoke.

"Hello. My name is George Morgan and I'm with American Energy Consultants[8] in California. We would like to order two liters of your part number 18593-0. Do you have that in stock?"

The clicking of computer terminal keys could be heard in the background.

"Okay, let's see, Mr. Morgan....that would be two liters of 18593-0......furfuryl alcohol. Correct?"

"Yes, that's correct. Could you ship immedi..."

"I'm afraid before we can ship that chemical to you Mr. Morgan we will need to know what you are using it for."

"Yes, of course. According to the engineering purchase request it will be used for acid catalyzed reactions."

"Could you elaborate more on that please," he probed.

"I'm afraid not," I parried. "I'm just a purchasing agent, not an engineer. Can you ship immediately by UPS?"

"What's your address?"

Fortunately, the nitric acid proved far less of a problem. When we ordered it from Corco Chemical the people there were excited and supportive of us in how we planned to use it. They took extra pains to make us a super strength batch. The nitric acid we ended up receiving from Corco was an impressive 98% solution. But the real story of the nitric acid was its delivery.

98% nitric acid is about the most powerful oxidizer you would ever not want to come in contact with. The label on the bottle, for example, warns that the contents will "dissolve human flesh." It also warns that inhalation of the fumes could result in illness or death. Because of the hazards involved, this chemical cannot be shipped by regular mail or UPS—only by common carrier (normal interstate trucking lines). Such carriers must display any and all hazard diamonds as

---

[8] AEC was a small company owned by my parents that did Title 24 energy audits for construction companies doing business in California. I didn't really have their permission to use their company name, but it sounded so good I knew it would help sell it.

required in the states the chemical will travel through, and each state has slightly different requirements as to which diamonds must be posted.

I had given Corco Chemical my insurance office address for the shipping destination. This was to make sure that someone would be available during normal business hours to accept the small 12-inch by 10-inch by 8-inch wooden packing box when it arrived.

One day I returned from a luncheon appointment to find the box sitting on the counter at the front of our office. It had been accepted and signed for by Mitch, one of the employees.

"When did this arrive?" I asked.

"About forty-five minutes ago," said Mitch. "A cowboy driving an eighteen-wheeler dropped it off."

"A cowboy?"

"Yeah, you know. One of those Okie-from-Muskogee types. And he was pretty pissed about something."

"About what?"

"Well, after he set the box down and we signed for it, he went out to his truck and tore off a bunch of stickers that had been plastered all over his rig. I could tell he didn't like having them on there."

"What kind of stickers?"

"They were all diamond shaped. A couple of white ones that were labeled "poison", several that were red that said "explosive", and a few yellow ones with the word "oxidizer"—whatever that means. It took him almost a half-hour to pull and scrape them all off."

"Did he say anything?"

"Only when he left."

"What did he say?"

"He just said, "Ah sure am glad to get that shit offa mah truck.""

T MINUS SIX MONTHS

Toward the latter part of 1986 we scheduled a solid rocket launch at Lucerne Dry Lake. The date for the launch would be January 31. This launch would serve two purposes; to have a fun group activity and, more important, to test the launcher that would be used later for the liquid rocket.

We built a 4-inch diameter, 6-1/2-foot-long solid rocket that was as close as possible to the size, shape and weight as the liquid rocket would have at takeoff. The launcher was a hinged T-rail design made of 2-inch square steel tubing. So simple was this launch rack that we bought the materials for it, finished the welding, and fully constructed it all in one day. Launch lugs, similar to the ones we would use on the liquid, were fashioned by Ron Milfeld and a successful fit-up was performed with the rocket and launcher.

On January 31, 1987 we met at Lucerne Dry Lake. The rocket was fueled with an old standard propellant combination: potassium nitrate and sugar. After firing several small motors to make sure our propellant mix was adequate, we mounted the large motor on the launcher, armed the ignition system, and retired to the firing area. The Lucerne Valley Fire Department, at our invitation, stood by with a fire truck.

The countdown went smoothly. The rocket ignited but chuffed for about forty-five seconds as it struggled to build up pressure. Then with a roar it sailed skyward at supersonic speed. We did not recover the rocket till weeks later, but it did not matter. The launcher had worked perfectly.

Upon returning home I called Dan (who had been unable to attend the launch) and related to him our success. Then I asked him a key question.

"What must be done to finish the nitric acid rocket?"

We made a list. Then, over the next few months, commenced knocking each item off it one by one. The nose cone was machined, the male pressure fitting was milled to its proper shape, the copper pressure feed lines were made. The valve plugs and valve restraining pins were touched up on the lathe and the Viton O-rings were purchased. The original injector with the broken drill bits was scrapped as unrepairable and a new one was machined (for free) by Bill Raybould at his machine shop in Los Angeles. When the final fin attachment was completed, we could see the rocket really taking shape.

So confident were we of our progress that in early April we set a launch date in late May. As time went on this proved to be too ambitious. As May came and went, we noticed our list of things to do was getting longer, not shorter. It quickly dawned on us that our original to-do list was woefully shortsighted. But the vision of the rocket ascending to heaven was starting to consume us. This rocket would fly, we decided, and it would fly soon. With a redoubled effort we reset our launch date for July 19, 1987.

Our group of rocket builders now consisted of Dan Ruttle, Ron Milfeld, his friend Paul Mcquown, my brother Stephen Morgan, and myself. We had been working together as a group for over a year and were starting to function like a real team. As the pace of construction quickened, our list of to-dos began to decline, and with the passing of June it looked like July 19 might really happen. Everything that was still unfinished were minor details, with one major exception: our innovative pyrotechnic valve ignition system had never been tested.

A CRITICAL SERIES OF TESTS

"One good test is worth a thousand expert opinions."

—Wernher von Braun

The pre-flight testing of the rocket consisted of nine tests: one ablative liner torch test, one hydrostatic test, and seven pyrotechnic valve tests. Following are the reasons for each test and their results:

THE ABLATIVE LINER TORCH TEST: Since the rocket engine would have no film or regenerative cooling, the ablative liner would have the responsibility of withstanding the heat and forces of combustion solely on its own. The linen-phenolic material we were using was untried and untested at these extremes and so we decided to subject it to an oxyacetylene torch to see what would happen.

During the construction of the engine, as you probably remember, we had cut off a two-inch section of the ablative liner. We now used this small piece for our first test. Since the theoretical burn time of the engine was seven seconds (see the hydrostatic test below) we torched the liner segment for exactly seven seconds. The material was charred but did not burn through. We repeated this procedure two more times with the same result.

Finally, we determined it would be advantageous to know the fail point of the material, so we subjected it to the oxyacetylene torch till it burned through. The fail time was ten seconds. We dubbed this first test a success.

THE HYDROSTATIC TEST: The purpose of this test was to verify that the propellant tanks would sustain a proof pressure slightly greater than the maximum expected operating pressure: MEOP.

MEOP for this rocket would be 300 psi and we decided to test them at a pressure 10% higher. The original design was for a tank pressure of 400 psi. Since 350 psi was the maximum delivery pressure of our regulator, we were forced to change MEOP to 300, since a 350 psi MEOP would not allow us to over-test. As a result of the lower operating pressure, we recalculated our theoretical burn time to be seven seconds.

With the tanks filled with water (a standard safety procedure) we connected the tanks to a nitrogen pressure bottle. We then cranked the regulator open to 50 psi. and waited. After about fifteen seconds we took the pressure up another 50 psi to 100. We continued in this manner, slowly stepping up the tank pressure. The reason for this procedure is so that in case the tanks rupture you have a much more accurate idea of what pressure they failed at than if you had cranked the pressure up quickly.

Moments later the tanks were pressurized to MEOP. Fifteen seconds later we took it up to 330 psi. The tanks held the pressure and there were no leaks; the external welds and both valves sealed perfectly. There was no way of knowing, however, if the internal welds between the two tanks faired the same. This test was solely to test the overall strength of the tank design.

The purpose of the next seven tests was twofold; to determine the amount of powder charge necessary to actuate the oxidizer and fuel valves and, once that was found, to insure that the system could be brought to a high degree of reliability. In each test the tanks were filled with water to simulate the propellant flow through the system and each test was performed at MEOP. All seven pyrotechnic valve tests were conducted by Steve and Dan at 8444 Melba.

PYROTECHNIC VALVE TEST #1:

In our first test the two valves were installed in their chambers and each was armed with a 5-grain black powder charge and an electric match as an igniter (all charges in these tests were fired with electric matches using FFFG black powder). The rocket was mounted vertically with the engine removed and the injector on. This was the standard setup for all the tests. The charges were fired with a car battery (the same energy source we would use for the launch). The charges fired with a mild poof. The valves did not actuate.

PYROTECHNIC VALVE TEST #2: After the rocket was disassembled and cleaned it was reassembled with the same setup but with 10 grain charges. When fired, the charges were louder, but the valves still did not actuate.

PYROTECHNIC VALVE TEST #3: Same setup but with 15 grain charges. This time when the charges fired the sound was so loud Steve and Dan were afraid a local neighbor might call the police on them. This time the oxidizer valve opened but not the fuel valve. Also, the oxidizer side's cover plate was bent outward by the force of the charge. This verified our suspicions that the cover plates, as they were originally designed, could be flawed. The bent cover plate could not be reused and another one had to be made (flawed though it was, it was all we had at the time). It was determined from this test that one of the causes of the valve failures was leakage of the powder charge gases through various ports, channels and holes in the valve block.

PYROTECHNIC VALVE TEST #4: The following steps were taken to rectify the problems from the previous tests:

(a) The cover plates were sealed using gasket material and electrical tape.

(b) a one-piece steel ring was machined to fit over the cover plates and strengthen them.

(c) the O-ring groove on the fuel valve was machined deeper to lessen friction and make actuation easier.

(d) the excess volume in the charge channels was reduced by machining wooden dowels to fit inside them. The charges were then placed in hollowed out cavities in the dowels. The maximum charge that would fit into each cavity was 7 grains. The same setup was used. The result, however, was the same as in the last test; the oxidizer valve actuated, the fuel did not.

PYROTECHNIC VALVE TEST #5: This time the fuel valve channel was cleaned and polished with emery cloth to a smooth finish, again for the purpose of reducing friction. This time the test was flawless. Using 7 grain charges, both valves actuated perfectly. The injector sprayed water all over the driveway at 8444 Melba.

PYROTECHNIC VALVE TEST #6: Same setup as test #5. Complete success.

PYROTECHNIC VALVE TEST #7: Same setup as test #5. Complete success.

During test #7 Dan took a close look at the spray pattern of the injector, getting quite wet in the process. All the streams seemed to be impinging correctly.

Both goals of these tests had been reached; a proper grain charge was determined, and the entire system was brought to a high degree of reliability.

FINAL ROCKET STATISTICS

Oxidizer	Nitric acid, 98% strength
Fuel	Furfuryl alcohol
Propellant mixture ratio	2/3
Tank design	Concentric
Tank pressure/gas	300 psi, Nitrogen
Pressure regulation	Blowdown
Chamber pressure	200 psi
Empty weight	38 lbs.
Takeoff weight	50 lbs.
Length	77 inches
Outside Diameter	4 inches
Fin design	4, 90-degree opposed
Chamber insulator	Linen-phenolic
Nozzle material	Graphite
Injector design	8 pairs of unlike doublets
Flight initiation	2 pyrotechnic valves, 12 vdc
Ignition system	Hypergolic paired propellants
Launch angle	85 degrees
Specific impulse	188 seconds (estimated)
Takeoff thrust	300 lbs. (estimated)
Burn time	7.30 seconds
Total flight time	80 seconds (estimated)
Maximum altitude	4 miles (estimated)
Range	.50 mile
Total expenditures	$2,500 (US)

ON TO NEVADA

It was 4:00 a.m., Thursday, July 16. The entire crew, with the exception of myself, had been up all night (I was going to be doing a lot of the driving) attending to the details of the rocket's final assembly. An important event had just taken place only moments before; the finished rocket was assembled and mounted, for the first time, on the launcher. It had been a long night. Ron, Paul, Steve and Dan stood for a silent moment admiring their work. There was no doubt about it now; this rocket would definitely be completed. Whether it would fly was the unspoken question on everyone's mind.

On the morning of July 18, 1987, Dan Ruttle, Stephen Morgan and myself assembled at 8444 Melba to do a final fit-up of the rocket to the launch rail, as well as take some last-minute photographs.

The rack was set up in the driveway. It was noted that several bolts were missing and so replacements were obtained. After the rack was assembled, we attempted to slide the rocket onto the T-rail via the launch lugs. The aft lug slid on perfectly, but to our stunned surprise the forward lug was way off. This problem was unexpected since a successful fit-up had been achieved Thursday morning at Ron's machine shop. After scratching our heads for a while Dan came up with the idea of reversing the forward launch lug, i.e. turning it around 180 degrees. This maneuver succeeded in resolving 50% of the misalignment, but not enough to allow the rocket to slide onto the rail easily. I suggested drilling the mounting holes of the male pressure fitting larger so as to allow more slop in the rocket's mounting. We then drilled the holes out half again bigger and remounted the fitting. Sliding on the rocket, we all breathed easier as both lugs went on in perfect alignment. To this day we do not know why the fit-up succeeded on Thursday and failed on Saturday.

The neighbors across the street graciously took several shots of us posing around the polished aluminum rocket, then we spent the next three hours packing Steve's station wagon for the trip. Not too soon, we were on the road (after a gas and Pepsi fill-up) and headed north on Highway 14.

Twenty miles outside of Mojave we passed the Ransburg turnoff where for so many years we had turned east towards the Mojave Test Area near Koehn Dry Lake. As we headed north, leaving the turnoff behind, I couldn't help thinking back to my high school years and the dozens of zinc/sulfur rockets flown by the Chaminade Rocket Club.

"Things sure have changed," I blurted aloud.
"What?" said Steve.
"Nothing."

Just before midnight we pulled into the parking lot of the Walker River Lodge in Bridgeport. Exhausted; we augered into bed with liquid rockets dancing in our heads, trying our best to get some sleep.

In the morning we crossed the highway to a small cafe and had a big breakfast. Afterward, we waited outside the motel room for the rest of the rocket crew. Before long, their cars came into view a few miles away and in a minute we were all together. Paul, Joanne, Ron, Joyce and her son, and Ted and his boy went across the street and had breakfast, then we were off for Gerlach, Nevada and the Smoke Creek Desert. Paul had brought walkie-talkies for every vehicle, so all four cars were able to keep in contact with each other as we caravanned northward. It was fortunate, too, because outside of Gardnerville Ted's car became separated from us, then, just before Reno, the same happened to Ron. With the advantage of radio communication, we were all able to stick together.

Soon we were passing through Reno, the "biggest little city in the world", and heading east on Hwy 80. We caught sight of the first 65 MPH signs we had seen in the last ten years.[9] At the last outpost before heading north to Gerlach we stopped for a food and drink fill-up. Now, on a little-used two-lane blacktop we started north toward the Smoke Creek Desert.

"We're almost there," I announced.
Then we passed a mileage sign: GERLACH 77 MILES
"Relatively speaking, of course."

The moon-like landscapes we now passed through were made even more eerie by the moving shadows of puffy clouds. We took turns pointing out other sights that looked good for rocket launches. Finally, over a small rise we went and there, between us and the horizon, were two small towns.

"The farther one is Gerlach," Dan mentioned, without explaining how he knew that.

We didn't suspect it at the time, but we were about to enter the most mis-named town in America.

---

[9] The Arab oil embargo of the mid-1970's had caused Congress and then-President Richard Nixon to lower the national speed limit to 55 mph as a fuel saving measure. The law was not repealed until 1995, but in the interim some rural states were allowed to bump their speed limits back to pre-embargo numbers.

BRUNOVILLE

Chuck Piper of the Rocket Research Institute had warned us, "When you get to Gerlach don't tell anyone who you are or why you're there. Be discreet and don't get noticed."

It was a logical request. Most people are interested in rockets until they find out you plan to fly them in their back yard. And though we would be thirty miles away, Gerlach was the nearest civilization and therefore the first potential line of offense in the event of a complaint. And so, we would be discreet, we decided. No one in Gerlach would know that we were in town or what we were there for. The RRI had been flying rockets out here for twenty-five years. We would not burn any of their bridges.

Past the U.S. Gypsum plant, past the town of Empire, we crossed the last few desolate miles to Gerlach. Entering the town, the first sign of life we saw was the Texaco station. Passing it slowly, we noticed a sign that read: BRUNO'S TEXACO. A few yards farther on we came across BRUNO'S MOTEL. Next door, BRUNO'S CASINO. Then BRUNO'S RESORT, BRUNO'S MARKET and BRUNO'S LIQUOR.

"Someone named Bruno throws a lot of weight in this town."

We made a brief stop at BRUNO'S PUBLIC TELEPHONE to notify the local air traffic control personnel of the impending launch. Hitting the edge of town, we put the pedal to the metal and roared out over BRUNO'S HIGHWAY into BRUNO'S DESERT. Overhead, BRUNO's SUN was heating up the day.

586 million years ago during the Paleozoic Era of Earth's history a large part of what is now Utah, Nevada and eastern California, today known as the Great Basin, was covered with water. It remained so for nearly 300 million years. In the Late Devonian Period large scale compression began to take place. Where there had been a shallow sea with a gently sloping undeformed bottom, mountain ranges and the central Great Basin itself began to rise. Just before the end of the Triassic Period the North American continent began a ponderous clockwise rotation. A period of volcanic activity and thermal bulging of the Great Basin followed.

As the continent began to break up and divide during the Jurassic, the highlands of the central basin uplifted, blocking many of the drainages to the west. Then, about 80 million years ago during the Late Cretaceous, another age of mountain building and upward thrusting occurred, this time along the eastern and southern borders of the Great Basin, cutting off all easterly drainage. The first blockage would one day become the Sierra Nevada. The second would become the Rocky Mountains. During the Cenozoic huge volcanic eruptions of white ash blanketed the Great Basin, turning it into a hellish environment. Within this enclosed basin, surrounded on all sides now by watershed divides, came runoff water from nearby glaciers, eventually forming a mammoth lake, Lake Lahontan. At its maximum height Lake Lahontan covered more than 8000 square miles. In the latter part of the Pleistocene Epoch, about 70,000 years ago, the northern glaciers retreated as the Earth began to warm. The level of Lake Lahontan fluctuated, then began to recede, until, about 8,000 years ago, mere remnants or "puddles" were left. Pyramid Lake in northwestern Nevada is the largest of these surviving puddles. Where Lake Lahontan completely dried up, large alkali sinks remained. One of the largest of these sinks is the Smoke Creek Desert.

We had only sketchy details regarding how to find the traditional Smoke Creek launch area. Over the last two decades various experimental rocket organizations had used this ancient primeval lakebed for firing hundreds of solid propellant rockets. It has been the scene of many a weekend rocket revelry over the years and many a person had traveled these roads before us. But we were new to the area and for all intent and purposes we may as well have been the first to cross this way, for every square mile looked like virgin untrampled desert wasteland.

But with a few scribbled directions we now set off across the blacktop, heading north and searching for a gravel road that would take us in a wide west/southwest loop around the lake bed's northern perimeter. Our inexperience showed as we were forced to backtrack twice in our attempt to find the road.

Once on it we sailed along at over 50 mph, our combined fleet of vehicles kicking up a voluminous cloud of dust. We were looking for a specific dirt road that would lead us to the dry lake bed's "shore" and to the area used for many previous rocket launches. After meandering around the desert for a half hour and not finding what we were looking for, we decided to take the next dirt road we found, follow it to the lakebed, and set up operations. This we did minutes later.

To our surprise it turned out to be the very road we were looking for, for as we arrived at the point where the hills and brush ended and the featureless moonscape began, the rocket related debris of those who had preceded us could be found within a wide radius. Their remnants were everywhere. An aluminum fin here, a battered nose cone there, a bent launch rail here, a length of wire there. Items valuable enough to be hauled to the middle of nowhere, but not worth hauling home. Someone wondered what archeologists would make of this area a thousand years hence. We could have spent a lot of time poking through all the junk, but we had a schedule to keep.

It was about 3:00 pm. With our FAA waiver expiring at.8:00 pm and with two solid rockets and one liquid rocket to fire we wasted no time getting started. Several members of our crew were at work setting up the launcher. Others were assembling the motors and readying the payloads. Within minutes of arriving we became a beehive of activity. I was busy assembling the launcher when I noticed out of the corner of my eye Dan standing quietly a short way out on the dry lakebed. He was gazing intently southward at the area that would be our down-range impact area. He seemed unusually quiet, almost like he was meditating. I decided to check up on him and walked over to join him. I followed his gaze over the gray-white surface, trying in vain to see what he was seeing.

"What is it, Dan?"
He paused for a thoughtful moment, then replied in a tone laced with worry.
"It's a sea of mud."

Which it was. For though the Smoke Creek Desert is technically a dry lake, it owns that title solely because there is no liquid water on its surface. However, since it is still the low point topographically for a wide area, what little rain and moisture do accumulate gravitates here. There may not be enough water to form a lake, but there is more than enough to form a vast inland sea of mud.

THE ROCKET'S RED GLARE

"There are one-thousand things that can happen when
you ignite a rocket. Only one of them is good."

—Thomas Mueller, rocket engineer

Ron, Paul and crew had brought two solid propellant rockets. These were fired first, each with varying degrees of success. The first, a single stage vehicle, blew out its forward bulkhead milliseconds after launch—about twenty feet above the launch rack. Pieces of debris scattered over a small radius with no real hazard to the onlookers. The second rocket, a two stage with movie cameras, performed significantly better. Both motors fired properly, providing a magnificent launch. Parachute system failure, however, resulted in both stages being demolished (along with the cameras and film).

It was unfortunate, as they had put a great deal of effort into their project. Silently to myself, I hoped it was not an omen of things to come.

With the second launch completed we commenced the job of setting up for the acid/alcohol rocket. The launch rail was lowered, and the main body section of the rocket was slid onto it, followed by the nosecone and its adaptor tube. The assembled rocket was then was raised into firing position.

We had decided previously that the back end of a pickup truck would be the perfect height for the raised propellant loading area we would need, and Ron's truck was chosen (since it was the only truck we had). After backing it up within a few inches of the rocket we commenced converting the truck's bed into our official launch gantry. A small table was set up and all the necessary tools were laid on it. The propellants, still protected by their packaging, were laid in separate corners. Finally, all unnecessary personnel were excused, and the loading procedure began.

As the chief project engineer, Dan felt personally responsible for the hazardous operation of fueling the rocket and so he volunteered for the task. Assisting him would be Steve, Ron and myself. Paul helped hook up the high-pressure nitrogen bottle to the launcher pressure fitting with fifty feet of copper tube. Then he assisted Steve with putting on the Scott air pack and showing him how to use it. It was 7:00 pm.

From this point on, we spoke only when it was necessary. Our FAA waiver would expire in precisely one hour—we had to work fast.

Dan Ruttle assembles the rocket on the launch rack.

Paul McQuown assists with attaching the pressure line.

Dan begins the fueling process by loading the furfuryl alcohol.

Paul McQuown assists Stephen Morgan with the Scott air pack equipment.

As Dan prepares to load the highly toxic nitric acid, Stephen stands by to assist in the event of an emergency,

Dan pours the nitric acid into the oxidizer tank. Note the protective Viton gloves.

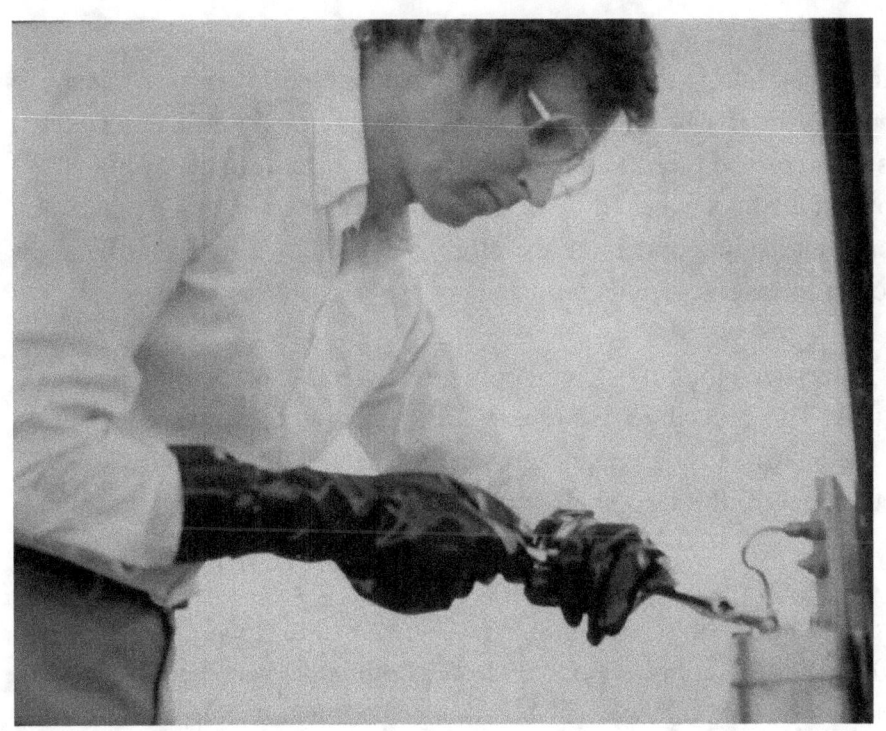

With the propellants loaded, Dan seals off the tanks.

The rocket—assembled, loaded, and ready.

Dan opened both alcohol tank ports, setting the plugs aside. He then put the small white plastic funnel into the fill port and made sure it was steady and secure. In preparation for this procedure we had brought two funnels—one for the fuel and one for the oxidizer. Because the chemicals were hypergolic, it was important that not even a drop of one come in contact with the other. As a safety measure for the propellant loading operation, therefore, our two funnels were different sizes and colors from each other in order to avoid confusion.

Taking a thoughtful breath, he unsealed the first alcohol bottle, removed the cap and then poured its contents slowly into the tank. This procedure was repeated with the second bottle, however only half of its volume was used. Next, a short copper tube was connected from one check valve to one of the alcohol ports. Capping off the second port completed the alcohol loading portion of the operation.

Ron's and my part in all this consisted mainly of getting everything on film. Ron ran the hand-held video camera while I clicked away on my Canon AE-l. We both lent a hand and occasional free advice as needed. Steve stood by with the Scott air-breathing equipment to help out in the event of any emergency.

Now it was time for the nitric acid. Weeks before, in preparation for this moment, Dan had taken an empty acid bottle, the same size and volume as we were about to use and filled it with water. Then, using a stopwatch, he timed how long it took to empty. The average emptying time, at a speed that did not cause splashing, was twenty seconds. And so, he had been practicing holding his breath for twenty-five seconds and more. Now this exercise would be put to the 'acid' test.

After removing both oxidizer tank ports, Dan peered into the tank through one opening while shining a flashlight through the other. This was to check for any leaks from the alcohol tank to its neighbor.

"I can't see anything," he announced, meaning the tank was dry and secure.

After a quick look around to assure himself there were no unnecessary personnel in the area, Dan began the job that he would later describe as, "Risking my life for a hobby."

"Steve," said Dan, "Hand me the Viton gloves."
"Where are they?"
Dan motioned to the station wagon.
"In the glove compartment."

With the Viton gloves on Dan placed the long blue funnel into the oxidizer tank. He then removed the acid bottle from its plastic bag container and placed it gingerly on the truck's bed. Unsealing its plastic cap, we all heard a distinctive hiss as a slight pressure was released. With the cap off, Dan grabbed the bottle firmly and raised it up to the funnel's lip.

I moved back.

And then he was pouring. The only sound you could hear was Steve's in-out breathing with the air pack. Twenty seconds later the acid bottle was empty and Dan cast it aside on the ground. One of the two ports was capped off followed by the second copper pressure feed line. Everything was going smoothly.

Removing the C-clamp that had been suspending the adaptor tube two feet above the rocket, Dan slid the tube and its nose cone down and began to fit it onto the forward bulkhead. But it refused to slide on. It acted as if it didn't fit. Dan pushed, shoved, turned it and cursed it, but the adaptor tube, which had fit perfectly so many times before, would not go on. I watched in disbelief as I realized that we had a major problem.

After a few moments watching Dan futilely struggle with the recalcitrant tube, I asked, "You want the rubber mallet?"

"I'm afraid so," He said reluctantly. "Yeah—hand me the mallet."

This was one tool we had not planned on using for this part of the operation. For that reason, the mallet was stowed in Steve's station wagon. I flew over to the car, grabbed the mallet, and ran back to the launch rack, almost throwing it into Dan's hand.

Wham! Wham! Dan started banging on the tube. I could see now what the problem was. The adaptor tube and the rocket body were off at different angles to each other, causing a misalignment. Then, with a stiff blow to the top of the tube, it slid down. We started breathing again. I looked at my watch: 7:45 pm.

"Damn holes don't line up," said Dan.
I looked up. Not another problem! "What's wrong?"

Contrary to my thoughts a few moments previous, our misalignment problem was not over. Of the four mounting bolt holes, only one would line up properly. Dan retrieved the discarded mallet and started banging again. But this time the problem resisted solution. Finally, muttering expletives under his breath, Dan tightened the lone bolt.

"It's as tight as I can possibly get it on," he reported.

And so the rocket and its forward adaptor tube would fly with only one lone bolt holding it all together. The experience reminded me of that great Wernher von Braun quote: "It takes 65,000 errors before one is qualified to build a rocket."

After the nose cone was installed, Steve armed the rocket by connecting the igniter wires to the long extension cord leading to the firing area. He and Ron then rejoined the rest of our party at the observation area. It was now Dan and me.

As Dan did a last-minute check of the rocket, I got into the driver's seat of the station wagon. The car had been parked next to the nitrogen bottle so that the regulator valve could be opened by someone reaching out the left side window of the rear seat. As Dan approached, I started the engine. He got in behind me and positioned himself next to the open window.

Reaching out, he grabbed the valve handle and began to slowly crank it open. We watched the pressure gauge creep up from 0 to 50 psi, then to 100 psi. Then 50... 200...250. . .

I caught myself crouching down in an instinctive attempt at self-protection.

Then the gauge his 300 psi.

"Okay," whispered Dan. "Let's get out of here."

We drove to the control area where Ron had set up his firing panel. Loren was waiting with the megaphone. As I double-checked my camera settings, Dan said to him, "Thirty seconds."

"THIRTY SECONDS!" came the amplified announcement.

A primordial quiet descended on the desert as the countdown continued. And then we were down to ten ... nine...eight...

"SEVEN...SIX...FIVE...FOUR...THREE...TWO...ONE..."

Dan pressed the firing button. A loud *POP* was heard as the pyrotechnic valves blew. Then, like the sound of an F-14 flying low overhead, the rocket's supersonic exhaust roared out of the nozzle. The rocket leaped quickly out of the launch rack, climbing high and fast. I clicked off one picture after another. The exhaust roar continued for the pre-calculated burn of seven seconds. Then, a mile above the desert floor, the propellants ran out. As the rocket coasted, it disappeared from view.

On terra firma, pandemonium was breaking out. Everyone, including (or especially) myself had lost all semblance of scientific composure as everyone danced, shouted, and high-fived each other. Still, I felt at a time like this it would be nice if someone said something appropriate for historical purposes. The right moments require the right words.

"Oh my god!" I practically screamed. "It was bitchin'!!"

Perhaps the rocket was still in flight. Perhaps it had already impacted somewhere out in that barren wasteland. Either way I didn't care, for my thoughts were on something else. Try as I might, I just could not get the picture of it out of my head. It was red, I thought to myself, pondering the line in our country's national anthem. The flame was red.

Lacking a zoom lens, this was the best takeoff shot we could get.

BORN TO DIG

We had been digging with pick and shovel for an hour. The sun had dropped below the horizon and so we were working in twilight now, taking turns. By 9:00 pm we had a hole four feet in diameter two feet deep; and we were exhausted. Finally, we stopped and set aside our shovels and held a pow-wow. Dan spoke first.

"Chuck Piper told me their rockets go into this goo as much as twenty-five feet down."
"If it's that far down," I said, "we'll never get it back."
"We have to get it back," said Steve. "We've put too much time, effort and money into it to just leave it here."

That was certainly true, but still one had to face reality. A twenty-foot hole would be a major excavation. We had neither the time, the equipment nor the personnel to carry off such an undertaking. Ron, Paul, Ted, and their families were leaving that evening to return home. That would leave only Dan, Steve and myself to do the excavation.

No way, I thought to myself, thinking I should have fought harder for the parachute system.

And so, everyone packed up and left. They to Reno and points beyond, we to Gerlach and Bruno's Motel. Steve, Dan, and I would stay overnight and try digging some more the next day. After all, we thought, it *could* be just below the surface. We had taken Monday and Tuesday off from work and so we had plenty of time on our hands. What did we have to lose by trying?

The ride back to Gerlach was uneventful. We checked into a room and met Bruno for the first time. He was running one of his many entrepreneurial affairs, Bruno's Casino, from which he also ran Bruno's Motel (and who knows what else).

"So, you are the famous Bruno," I announced. "Looks like you practically own this town."
"Yeah," he replied. "Wanna buy it? Write me a check."

We had dinner at, yes-you-guessed-it, Bruno's Restaurant, then played a few of our quarters in Bruno's slots. George's quarters quickly became Bruno's quarters. We talked of the day's events and how spectacular the launch had been, and of our plans for the next day. We were all pessimistic about our chances, but somehow someway, we decided, we would get that rocket back. Even if we had to dig down twenty feet with our bare hands.

We slept well that night, and why not; Gerlach-Brunoville is a very quiet town. When I awoke, I noticed Dan was still asleep, but that Steve was gone. I looked at my watch: it was 6:00 am.

The July morning sun had already turned night into full daylight. Slipping on my clothes I left to find Steve. It didn't take long—I found him pacing excitedly in the lobby.

"George, you're awake! Come here—I gotta show you something."

I didn't share his excitement, especially at six in the morning. Especially before breakfast. Especially in Gerlach-Brunoville.

"I found something that we can use to get the rocket back," he said.
I started waking up. "What!?" It was a better waker-upper than a bucket of ice water.
"A backhoe. The gas station next door has a backhoe!"

Not wanting to appear as ignorant as I was, I purposely failed to ask the obvious question, i.e. what is a backhoe? (Heavy machinery has never been one of my strong suits.) So, I silently followed him outside and toward the station, which was right next door.

"There," he said, pointing to the station's back lot. "There it is. What do you think?"

Approximately fifty feet behind the gas station was this massive yellow machine. In the front was a huge bulldozer scoop large enough to carry a Volkswagen around in. Connected to it were two mammoth hydraulic actuators, which protruded from the center of the beast, above which was the control area. There a covered cab contained an operator's seat and a myriad of control levers and pedals. Behind this was a long, powerful arm, bent in the center like the neck of a brontosaurus, where it was elbowed and again connected with hydraulic rods. At the end of this section rested a smaller, but sharper, scoop capable of gouging hundreds of pounds of earth and mud in a single sweep. In short, it was a machine that was born to dig.

And it was just what we needed.

LUCK IN THE MUCK

The smell of bacon wafted through the desert air. On the Coleman stove scrambled eggs were quickly frying. Soon we would be having a home-cooked meal out in the middle of nowhere.

We had arrived early from Gerlach that Monday morning to make one more attempt to dig the rocket out of its muddy grave before taking the expensive step of renting the backhoe. As I stirred the eggs in the pan I thought back to the events of that morning.

After Steve had pointed out Bruno's Backhoe, we had woken Dan to show him. Together we met with Cecil (Bruno's son-in-law and the manager of the service station).

"How much to rent your backhoe?" Steve had asked.
"Fifty dollars per hour, which includes the operator."

We held a meeting and quickly decided to accept his offer. Then Dan thought to ask him a pertinent question.

"Do you charge for the transportation time back and forth from the excavation site?"
"Of course," said Cecil.
"How fast does that thing travel?"
"About fifteen to twenty miles per hour."

A few swift mental calculations revealed that it would cost us almost $200 just for transportation costs to and from the launch site. Then there was the time needed for excavation. The final bill could tally hundreds of dollars.

"We'll get back to you," we said.

By now the entire town of Gerlach was abuzz with the news that a liquid fuel rocket had been flown the previous day on the nearby dry lakebed. Somehow the word had got out. Everywhere we went people would say, "Aren't you those rocket guys?" When this happened that morning while having hot chocolate in Bruno's Restaurant, I asked Dan, "Didn't Chuck Piper ask us to be discreet?"

Back on the lakebed, my mind turned to finishing breakfast. As we scarfed down bacon, eggs, and toast, we began to discuss the job at hand. Analytical minds went to work. The day before we had removed enough cubic yards of earth to leave a two-foot deep, four-foot diameter hole.

It had taken six people working two shifts about an hour in the thin air[10] to accomplish that, and we tired easily.

The hole we would eventually need would require removing possibly twenty times that much volume and we would have half as many people to do it. After a few more minutes staring at the hole, we headed back to town. We found Cecil in the service bay doing some welding on a go-cart.

"Do you take credit cards?" I asked.
"Sure do," he said, smiling like a factory worker on payday.
Unfortunately, I was the only one who had brought a credit card. I pulled it out of my wallet and handed it over.
"I guess you got a deal."

A couple of hours later we were back at the launch site awaiting the backhoe. Upon arriving there we noticed our left rear tire was flat. Using the pressurized nitrogen from the K-bottle, we jury-rigged McGyver-like a device with which we were able to connect the nitrogen bottle to the stem valve and re-inflate the tire. It took quite a bit of time, but as we finished, we looked up the desert road and saw the yellow monster coming our way, a cloud of dust in its wake.

A moment later we were congregated around the impact area. Along with Cecil had come his employee, Ned, and Cecil's son Willy. We outlined an area that we wanted him to dig in which would excavate an area immediately alongside the rocket's path of travel. With a wave of a hand and an "Okay," he began.

Moving levers and pedals in harmony, Cecil gouged the backhoe scoop deep into the dry lakebed. Moments later, the claw-like device brought up several hundred pounds of mud and clay. The machine maneuvered its cargo to an area off to its right, stopped, and attempted to dump its load by turning the scoop upside down. To our amazement, the entire mass of slime stuck like glue to the scoop. It refused to drop.

Cecil began to rock the scoop up and down, back and forth. Finally, with a distinctive sucking sound, the wet goo slid out and plopped onto the ground.

Dan, Steve and I exchanged glances. We could see this was going to be a slow process. And a slow process meant an expensive one—I could almost feel the dollar bills, hundreds of them, flying from my pocket. Dan looked like he was feeling the same way.

After a short while someone shouted, "Stop!"

---

[10] The Smoke Creek Desert lies at an altitude of almost 4,000 feet.

Looking into the hole we could all see the gleam of polished aluminum. Reaching in with a shovel, Steve brought out a piece from one of the smoke flares.

We were on the right track.

"Keep digging!" shouted Dan.

The backhoe resumed its labors. Cecil was starting to develop a technique for dumping the mud quickly and so the pace was picking up. At the eight-foot level we spotted another piece of aluminum, but much bigger. It turned out to be the forward adaptor, split along one side and now more resembling a flat piece of sheet metal than a tube.

"Keep digging!"

At this point we decided to widen the hole another two feet and dig directly over the impact point. Soon the new hole was even with the first and more debris was found.

"Keep digging!"

And then we were at thirteen feet—the depth limit of the backhoe. And still no rocket. With Steve lowered into the hole he poked around with a metal pole and announced that he felt something solid.

"It's about another two feet down."

"Keep digging!"

Before long we had a rectangular hole thirteen feet deep, five feet wide, ten feet across. We lowered Steve to the point precisely above where the rocket should be. Prodding around with a four-foot pole, he hit something solid. Way above him we all heard it. Then, looking up, he announced, "This is it!"

I had my doubts, and so did Dan. We had been hitting various pieces of metal for the last two hours. Whatever Steve had hit was only a few feet down, but it could have been anything.

"This is the tenth time we thought we'd found it," I said to Dan. "It' s probably just another smoke flare."

Down below, Steve could hear our apprehension.

"This is *it!* I'm telling you I've found it!"

There was only room for one person to dig at the bottom of the hole. Remembering our experience from the day before two feet may as well have been a mile.

Dan looked up at Cecil. "Is there any way you could get us another couple of feet?"

Cecil paused for a moment, then nodded. Raising the scoop, he brought Steve back up to the surface, and then commenced a clever solution to the problem. Using the bulldozer scoop he cleared away a large area, two feet deep, beside the hole. Then he drove the backhoe into the new depression. This effectively increased the digging machine's depth limit another two feet. Brilliant.

With renewed energy and enthusiasm Cecil began hauling huge shovelfuls of Smoke Creek mud to the surface. Man and machine appeared to be working in a new unison, and in only a few minutes our massive excavation was two feet deeper.

We lowered Steve into the hole again. Pole in hand, he began poking into the mud. Immediately we heard a loud *thunk*. We lowered a shovel down to him, and as he took his first scoop, he said, "I'll need a bucket to throw the mud into."

I rushed over to Ned's van, retrieved a five-gallon bucket, and returned to the hole.

"I've got the bucket," I announced.
"Forget it," said Steve. "Don't need it after all."

Below in the muck I could see Steve had found a rather novel solution. As he brought up a shovel full of mud, he would simply heave it against the wall of the hole, where it would stick like instant-setting glue. No bucket would be needed.

Clunk!

Straining our eyes downward, we could see Steve had discarded the shovel and was digging by hand. As he pushed the mud aside to the left and right, we saw it. There, clear as day, were the four opposing lines, 90 degrees to each other, that were the rear edges of the four fins. Also visible was the nozzle of the engine.

Dan and I smiled at each other. Ned and Willy smiled at each other. Cecil, at $50 an hour, took a break.

Ned grabbed a chain from his van and lowered one end down to Steve. Meanwhile, Steve was poking long holes into the mud all around the rocket body for air passages. This was to prevent the rocket from pulling a vacuum in the wet goo as it was lifted out, which might have been strong enough to prevent the rocket's removal. Finishing that, Steve attached the chain's hook around the rear launch lug as we wrapped the other end around the backhoe scoop. In a moment we were ready. Steve was brought to the surface and we motioned to Cecil.

"Okay! Pull 'er up!"

Even with the air passages to help out, the scoop had a tough time getting the rocket started up. We could hear a distinct sucking sound as the Earth attempted to retain its prize. Then suddenly there it was—dangling in the air, swinging free back and forth.

"Up! Up! Bring it up!"

The scoop could not raise any higher, so we all grabbed onto the chain and pulled it up by hand. Seconds later the rocket was resting atop the surface of the lakebed.

And then the acrid smell of nitric acid hit our nostrils. Apparently not all the acid had been burned in the rocket's combustion chamber—some small residual amount had survived, and now we were breathing it. Yet another unexpected event in a long line of unexpected events. We set the rocket down and backed off to get some air.

Eventually the wind blew off the fumes and we deemed the rocket safe to handle. Setting it in the backhoe scoop, Cecil drove it back to our car.

Cecil digs down more than 20 feet to reach the aft end of the rocket

EPILOGUE

It was Sunday, September 6th, 1987. Six weeks had passed since the launch of the Pacific Rocket Society's acid/alcohol rocket. Gerlach and the desert, having returned to normal for a short while, was receiving visitors.

Bill Wood, a member of the experimental rocketry group Starflight, looked up and gazed in dismay at the hazy skies. The fires of northern California were belching out millions of tons of ash and dumping them over tens of thousands of square miles. Here, to the Smoke Creek Desert, 150 miles away and seemingly out of range such calamities, easterly winds were carrying an unending stream of this pollution.

Members of Starflight were here to launch two Ammonium perchlorate and HTPB rockets. The smokey skies bode ill for tracking rockets that would attain altitudes in excess of 20,000 feet. The visual ceiling this day would be less than half of that amount. As the crew went through their preparations, Bill decided to examine the remnants of the now famous liquid rocket excavation site just south of them. A few minutes and a short walk later, he was standing beside the large mound of dried mud and clay.

The pile of loose earth rose incongruously above the flat dry lakebed about five feet high. Amazed at the extent of the excavation, Bill walked slowly around the pile, trying to imagine the size, depth and shape of the original hole that had been dug here.

Suddenly something caught his eye—something that didn't belong.

Stooping over, he picked up a small, twisted piece of aluminum from the middle of the mud-pile. It was rounded on one side with two welded pieces facing each other on the other. The metal had been traumatized by a severe impact, and one end of it was seriously bent over. Even so, its identity and purpose were unmistakable. Bill smiled as he realized he had just beaten one in a million odds. By some strange twist of fate, he had stumbled upon the forward launch lug from the PRS acid/alcohol rocket. It had just been lying there atop the excavation pile where anyone could have seen it. It could have ended up anywhere—the five-foot level, the ten-foot level, the twenty-two-foot level. But there it was—on top of the pile in plain view.

Pocketing it, Bill strode back to help with the firing.

PROJECT PARTICIPATION LIST

Chief Project Engineer   Dan Ruttle
Rocket and Launcher Design: Dan Ruttle

Rocket Fabrication:
  Nose cone   George Morgan
  Forward adaptor, propellant tanks,   Dave Griffith, Dan Ruttle
    pyrotechnic valve system
  Cover plate/restraining ring   Stephen Morgan
  Injector   Bill Raybould
  Combustion chamber, ablative   George Morgan
    liner and nozzle
  Fins   Dave Griffith
  Fin brackets and assembly   Bill Mensing, George Morgan
  Launch lugs   Ron Milfeld
  Launcher fabrication   Stephen Morgan, Paul McQuown,
                                                      George Morgan

  Firing panel   Ron Milfeld

System Tests:
  Ablative liner torch test   Stephen Morgan, Paul McQuown,
                                                      Dan Ruttle

  Hydro static test   George Morgan, Stephen Morgan,
                                                      Dan Ruttle

  Pyrotechnic valve tests   Dan Ruttle, Stephen Morgan
  Pyrotechnic cartridge and   Stephen Morgan
    cover plate modifications
  Rocket launcher fit-up   Ron Milfeld, Stephen Morgan,
                                                      Dan Ruttle

Launch Operations:
  FAA coordination   George Morgan
  Launcher setup   Loren Martin, Paul McQuown,
                                                      Stephen Morgan

  Safety crew   Stephen Morgan, Paul McQuown
  Propellant loading operator   Dan Ruttle
  Photography   Ron Milfeld, George Morgan,
                                                      Joanne Buffa

  Countdown announcer   Loren Martin
  Firing panel operator   Dan Ruttle

Recovery Operation: Paul McQuown, Stephen Morgan,
                                         Dan Ruttle, George Morgan,
                                         Cecil and Willy Courtney, Ned Waldner

Project Documentation:
    Videographer   Ron Milfeld
    Video editor   Craig Rosevere
    Video producer   George Morgan
    Project report   George Morgan

Engineering drawings: Dan Ruttle, Stephen Morgan
                                        George Morgan

ACKNOWLEDGEMENTS

First, I would like to thank my parents Richard and Mary Morgan who allowed us to use their West Hills home (referred to in the report as 8444 Melba) for the construction and assembly of this project.

Next, I would like to acknowledge all the members of the Pacific Rocket Society who teamed up to help us make this project a flying reality. I am no longer active in the PRS, but as of the date of the publication of this third edition (September 2025) of the acid/alcohol rocket report, the Pacific Rocket Society can be contacted through the following web site: http://www.translunar.org/prs/

Bill Mensing at Sheetcraft in Santa Paula, California, was indispensable in his role as fin assembly welder.

The man and his company, whose names we have forgotten, that did the crucial propellant tank welds back in 1980. The overall safety of the project depended more on his expertise than anyone else's.

Corco Chemical for being so helpful in making us a super strength custom batch of nitric acid.

American Energy Consultants, Canoga Park[11], California, for the usage of their Robocad computer drafting equipment and their assistance in financing our lathe.

Bill Raybould at W.R. Machine, El Monte, California, for making our replacement injector in record time and exactly to specs.

Trick Racing Products, El Monte, California, for the use of their shop, equipment, and general assistance.

Dave Griffith, Aerospace and Industrial Machining Co., for the use of his shop, equipment and moral support.

Cecil, Ned and Willy for their help in retrieving the rocket from the Smoke Creek Desert mud, and their never-say-die attitude.

The lady at Valley Seal who made sure the Viton O-rings were really Viton.

Air & Space Magazine and the Smithsonian Institution for allowing us to advertise this report.

---

[11] Later renamed West Hills

Craig Rosevere for his excellent job of editing our video.

The Holiday Inn, Victorville, California, for their assistance with our Oct. 3, 1987 seminar.

Bill Wood for giving us coverage in his Rocket Newsletter.

And anyone else I may have left out.

Thank you one and all!

TECHNICAL APPENDIX

DESIGN CALCULATIONS ON THE ACID/ALCOHOL ROCKET MOTOR

Since the outside diameter of the rocket was 4 inches, and since the injector would be bolted directly to the valve block, the maximum outside diameter of the combustion chamber was limited to the space needed for a circular row of mounting bolts. This maximum O. D. turned out to be exactly 3 inches.

To design the engine, we needed a value for L*. Since we did not plan to static test the engine prior to launch (organizing a static test would violate the Keep It Simple rule) we needed to determine a value for L* through other than experimental means.

Students of rocket history should be aware of the evolution of L*. In the early days, as a safety factor, rocket engineers used unnecessarily large L* values. As time went on and their experience increased, L* values declined, sometimes dramatically. The Bell Xl, for example, was originally designed with an L* of 24 inches. Later, in one abrupt change, the L* was shortened to 9 inches, a decrease of over 60%, with no drop in performance!

We decided to use the experience of the Bell Xl in obtaining our L*. The WAC Corporal, a ballistic missile contemporary of the Bell Xl (and therefore its technological cousin) used similar propellant s and injector design as ours. The L* used for the WAC Corporal was 73 inches. The after/before ratio of the Bell Xl L* was 9/24 or .375. Multiplying 73 inches times .375 gives a value of 27.4. We then settled on 28 inches as our L* value.

Arbitrary? Unscientific? Maybe. But it worked.[12]

To determine the area of the throat we use:

$$F = Cp \times Pc \times At$$

where F equals engine thrust in pounds, Cp equals the coefficient of thrust, Pc equals the chamber pressure and At equals the area of the throat. The value of F was preset at 300 pounds, the value of Pc at 200 pounds. From Sutton's tables we obtain the value of Cp as 1.27.

This gives us: $300 = 1.27 \times 200 \times At$

$At = 1.181$

---

[12] Such guesswork was actually quite common in the early days of rocketry and, sometimes, still occurs.

Knowing the area, we can calculate the diameter, giving us a throat diameter of 1.227 inches.

We can now calculate the volume of the chamber by:
$L* = Vc/At$ where Vc equals the volume of the chamber. This formula then gives us:

$28 = Vc/1.181$

$Vc = 33.07$ cubic inches

Since the inside diameter of the chamber was preset by several factors (thickness of the chamber wall and ablative liner) at 2.15 inches, we would need a chamber length of about seven inches to obtain this volume. Since the chamber tube we had already purchased would give us a chamber length of nine inches, we decided to use it 'as is' and enjoy a slight reliability factor.

Back to the WAC Corporal.

The injector on that rocket used eight pairs of unlike doublets in its injector design. We liked the simplicity of the design and so we adopted it. The size of the injector holes, however, would be much smaller. We used in our calculations a pressure drop across the injector of 80 psi and a discharge coefficient of .75 (see example 9-1 in George Sutton's 3rd edition).

For an ISP (specific impulse) value we used that obtained by the Aerobee missile using the same propellant combination: 188 seconds. The total flow rate of both propellants combined is found by:

Flow Rate = F/ISP = 300/188

giving us a total flow rate of 1.596 pounds per second.

The mixture ratio of acid/alcohol is 2/(2 + 1) = 2/3 (from George Sutton's first edition).

Therefore, the oxidizer flow = 2/3 x 1.596 = 1.064 lbs./sec, and the fuel flow will be 1/3 x 1.596 = .532 lbs./sec.

Since we have eight oxidizer holes and eight fuel holes, the propellant flow through each individual hole is found by:

Flow through 1 oxidizer hole = 1.064/8 = .133 lbs./sec.
Flow through 1 fuel hole = .532/8 = .067 lbs. / sec.

To obtain the cross-sectional area of each hole we need the densities of each propellant. From tables we find them to be:

9 lbs./cubic foot for nitric acid and 70.5 lbs./cubic foot for furfuryl alcohol.

From standard formulas contained in Sutton's book we calculate the area of the oxidizer holes by:

$$\frac{.133 \times 144}{.75\sqrt{2(32.2)(93)(80)(144)}} = .003 \text{ in. sq.}$$

and the area of the fuel holes by:

$$\frac{.067 \times 144}{.75\sqrt{2(32.2)(70.5)(80)(144)}} = .0018 \text{ in. sq.}$$

Knowing the areas of the holes we now calculate their diameters to be:

Oxidizer hole = .062 in.
Fuel hole = .048 in.

These are precisely the size holes which were then drilled in the injector.

BIBLIOGRAPHY

Sutton, George P., <u>Rocket Propulsion Elements</u>, John Wiley & Sons, Inc., New York, 1967; 3rd edition.

Sutton, George P., <u>Rocket Propulsion Elements</u>, John Wiley & Sons, Inc., New York, 1967; 2nd edition.

Sutton, George P., <u>Rocket Propulsion Elements</u>, John Wiley & Sons, Inc., New York, 1967; 1st edition.

Huzel, Dieter K. and David H. Huang, <u>Design of Liquid Propellant Rocket Engines</u>, Rocketdyne Division, North American Rockwell, Inc., NASA SP-125, Washington D.C., 1971.

<u>An Introduction to Rocket Missile Propulsion</u> (author uncredited), Rocketdyne Division, North American Rockwell, Inc., 1958.

Fiero, Bill, <u>Geology of the Great Basin</u>, University of Reno Press, 1986.

Wheeler, Sessions S., <u>Pyramid Lake, Nevada</u>, Caxton Printers, 1967.

Clark, John D., <u>Ignition! An Informal History of Rocket Propellants</u>, Rutgers University Press, 1972.

Kit, Boris and Douglas S. Evered, <u>Rocket Propellant Handbook</u>, New York, The MacMillan Company, 1960.

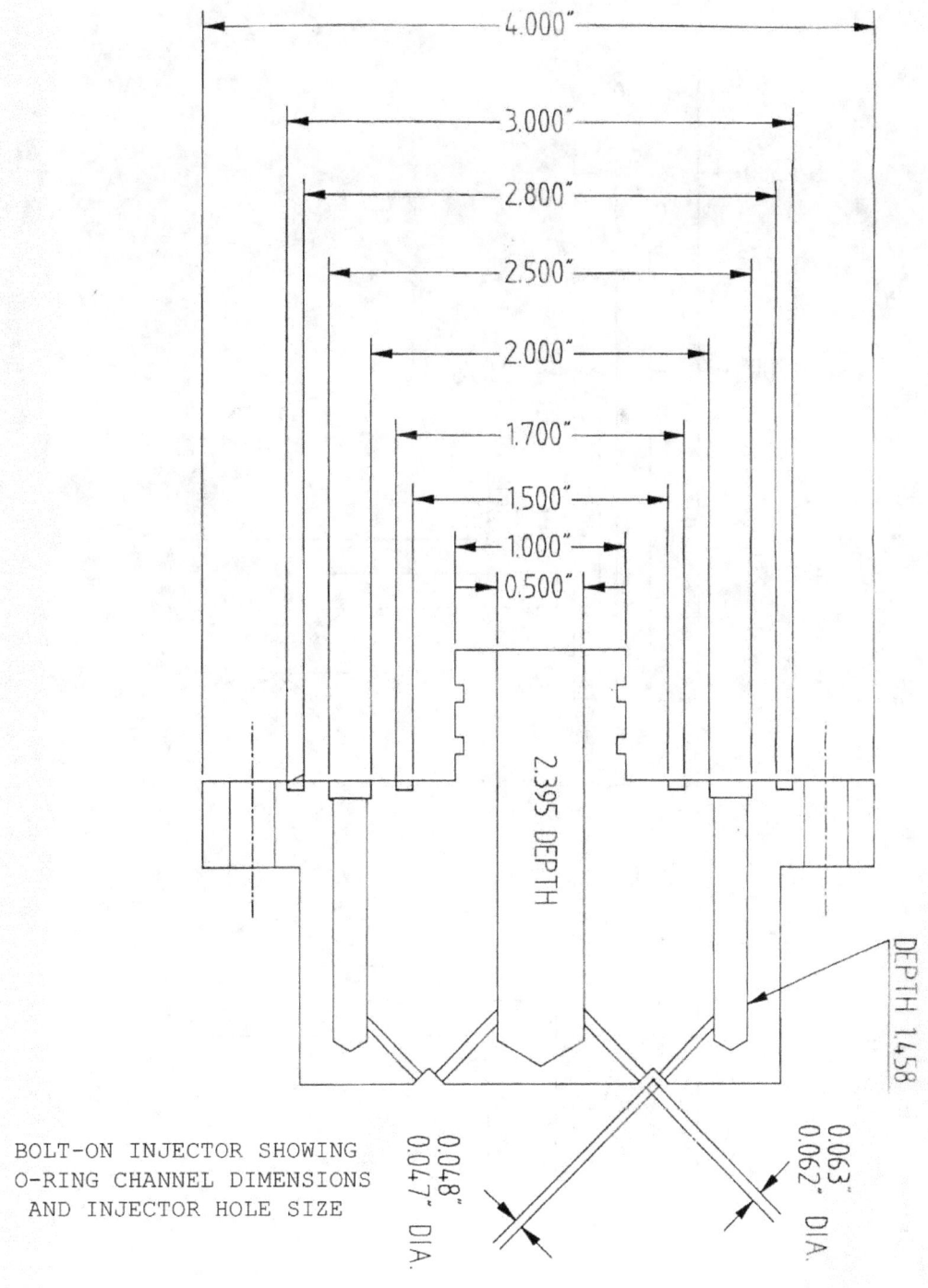

4.000"
3.000"
2.800"
2.500"
2.000"
1.700"
1.500"
1.000"
0.500"

2.395 DEPTH

DEPTH 1.458

0.048"
0.047" DIA.

0.063"
0.062" DIA.

BOLT-ON INJECTOR SHOWING
O-RING CHANNEL DIMENSIONS
AND INJECTOR HOLE SIZE

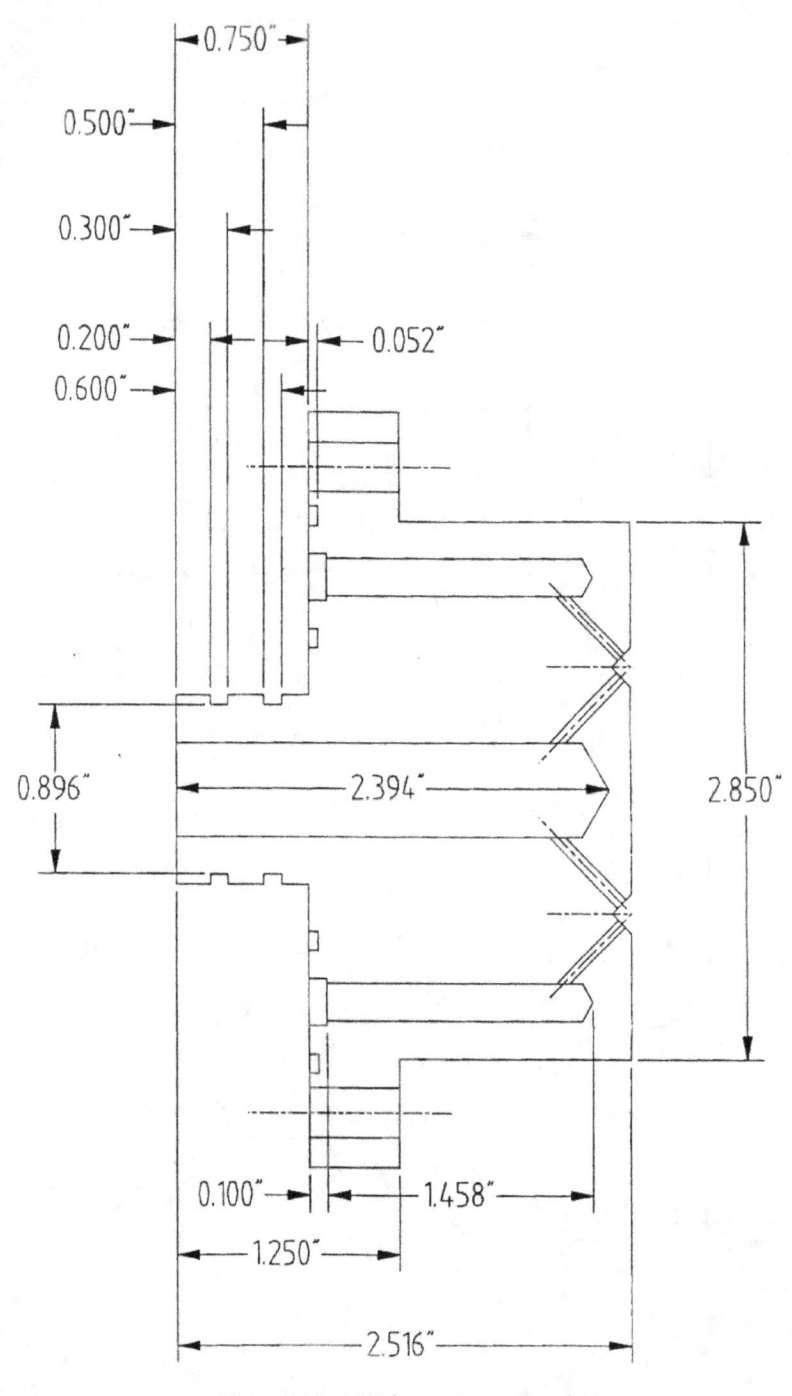

CUTAWAY OF ACID/ALCOHOL INJECTOR

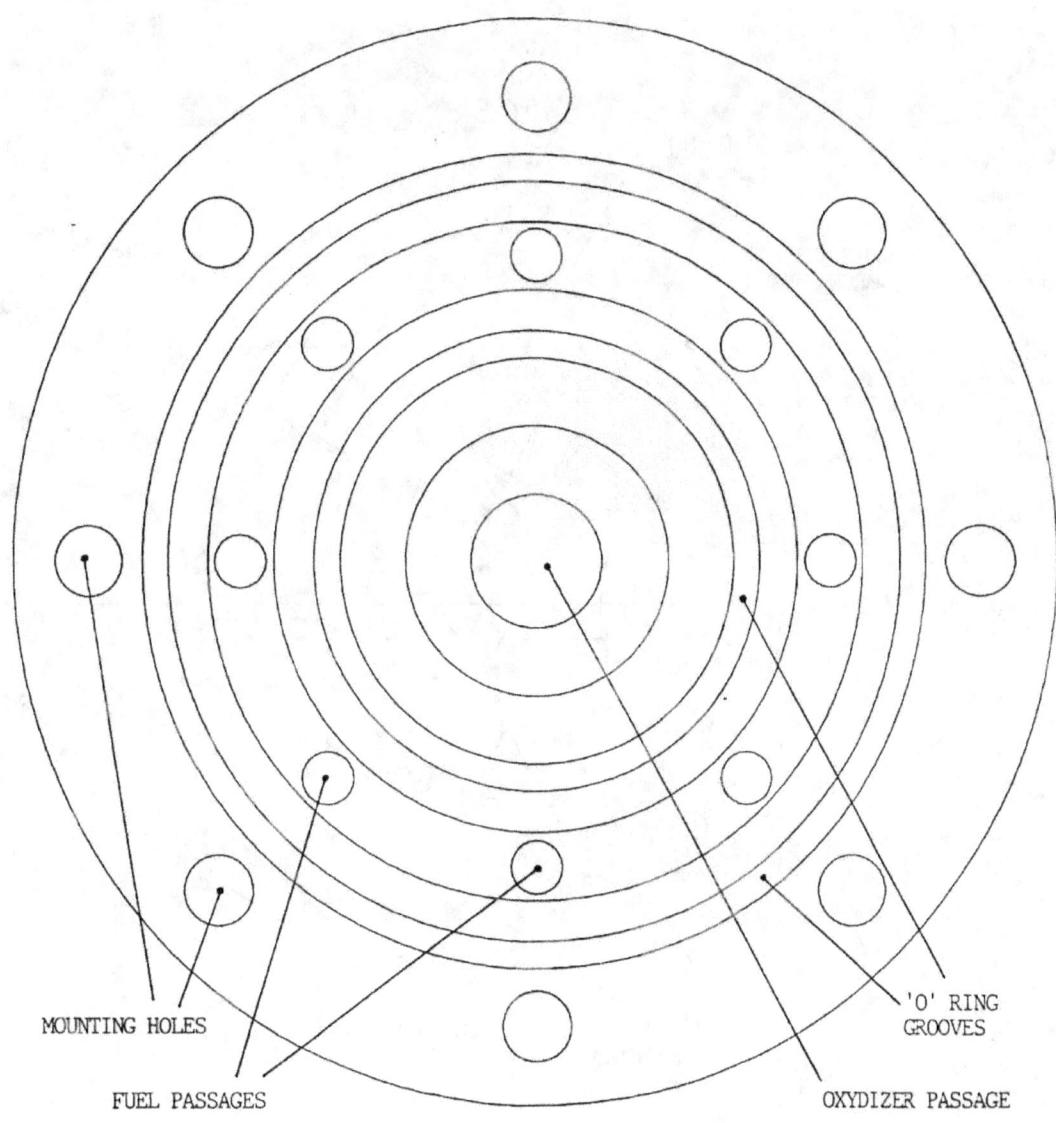

MOUNTING HOLES

FUEL PASSAGES

'O' RING
GROOVES

OXYDIZER PASSAGE

TOP VIEW OF ACID/ALCOHOL INJECTOR

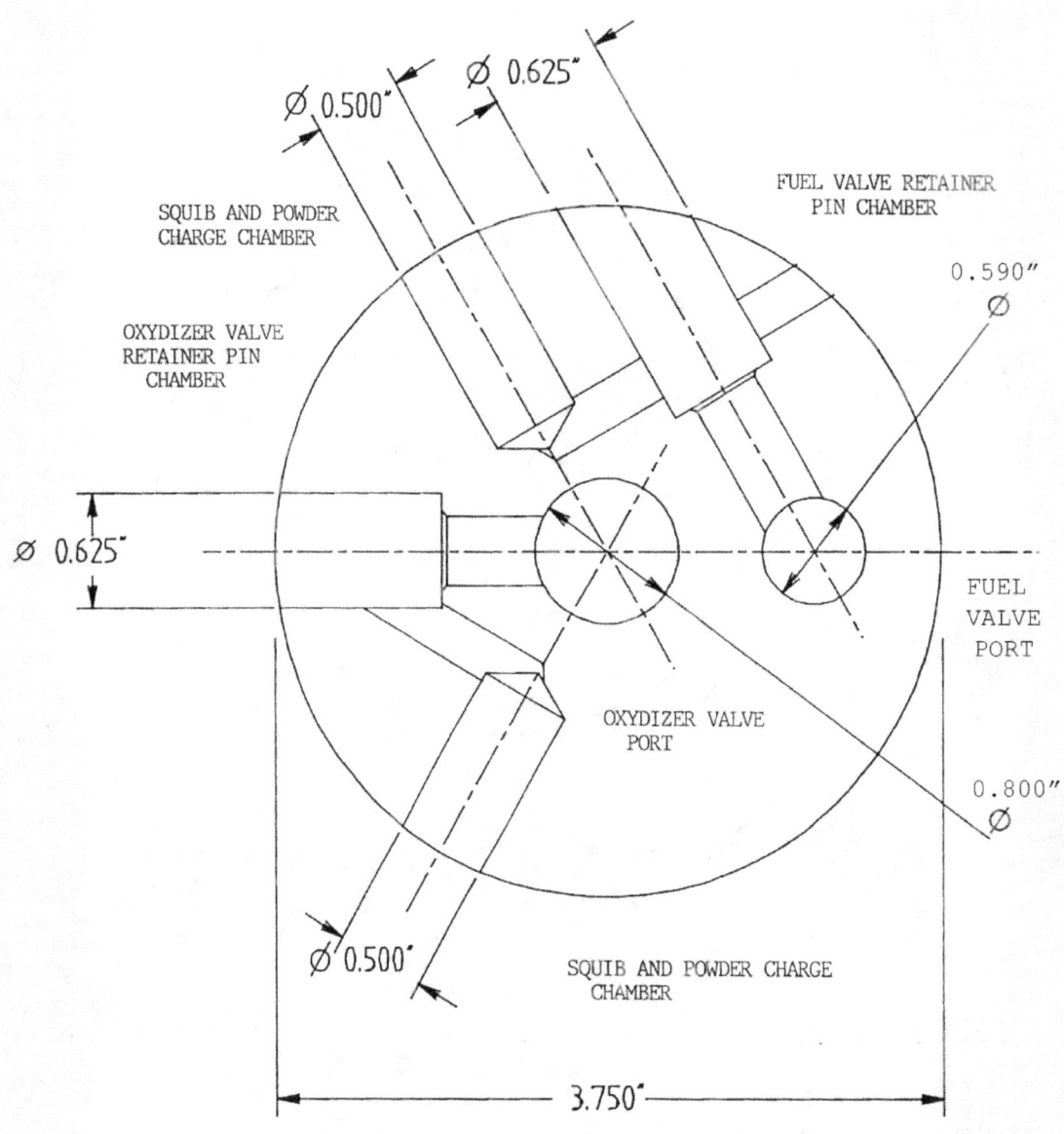

Ø 0.500″

Ø 0.625″

FUEL VALVE RETAINER
PIN CHAMBER

SQUIB AND POWDER
CHARGE CHAMBER

0.590″
Ø

OXYDIZER VALVE
RETAINER PIN
CHAMBER

Ø 0.625″

FUEL
VALVE
PORT

OXYDIZER VALVE
PORT

0.800″
Ø

Ø 0.500″

SQUIB AND POWDER CHARGE
CHAMBER

3.750″

CUTAWAY VIEW OF PYROTECHNIC VALVE BLOCK, PINS AND VALVES REMOVED

68

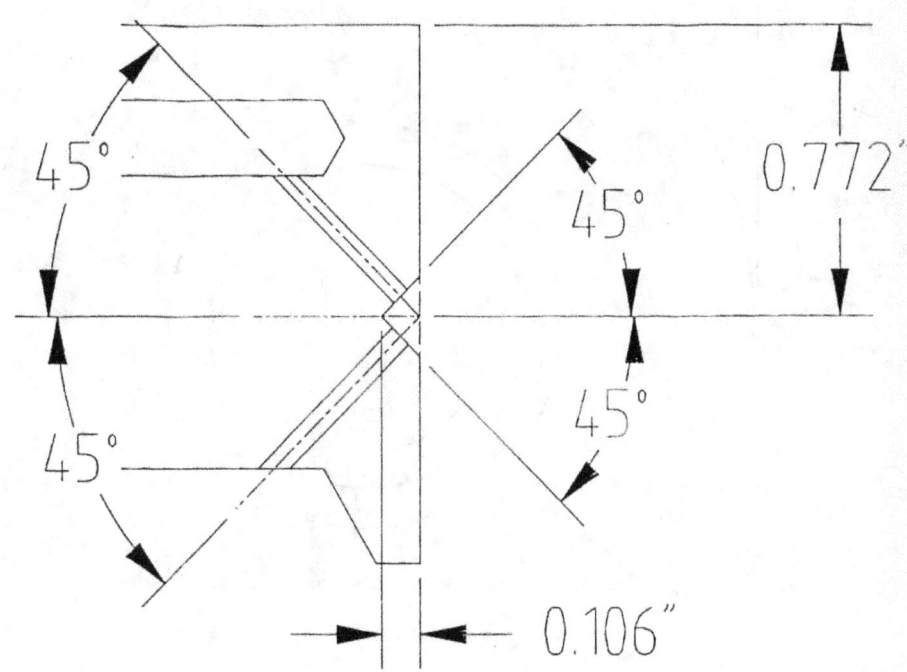

DIAGRAM SHOWING INJECTOR HOLE PLACEMENT AND ANGLES

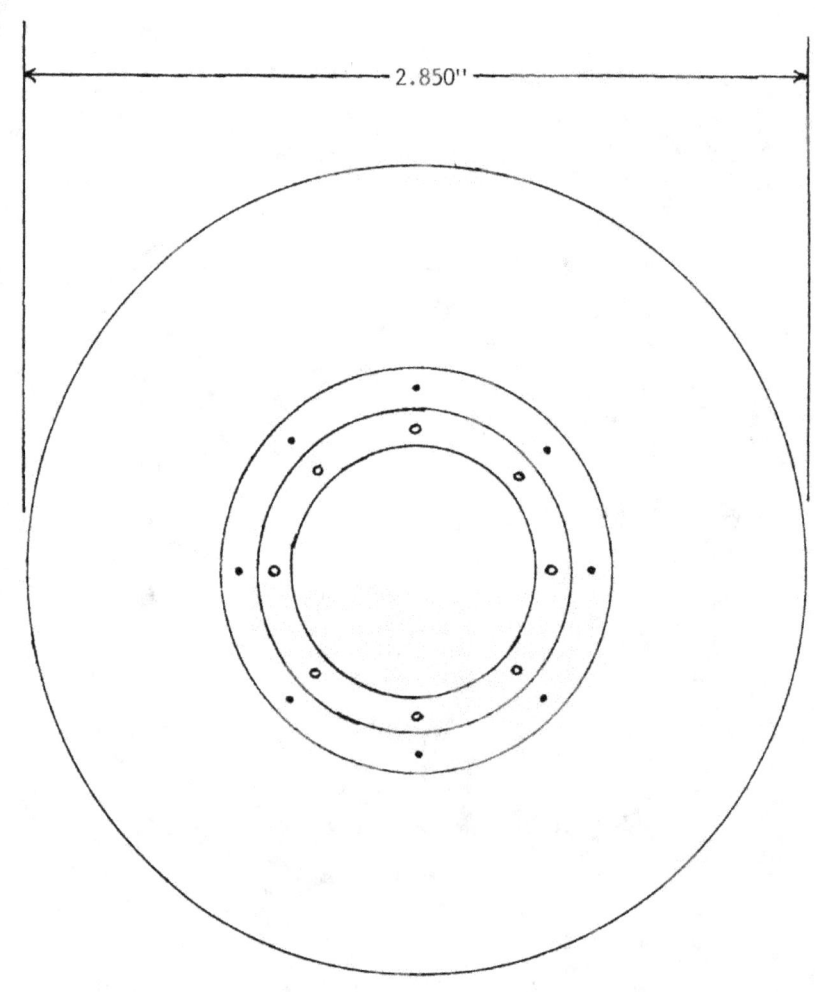

2.850"

INJECTOR FACE SHOWING PLACEMENT OF THE EIGHT PAIRS OF
UNLIKE DOUBLETS.   FUEL HOLES ARE SHADED.

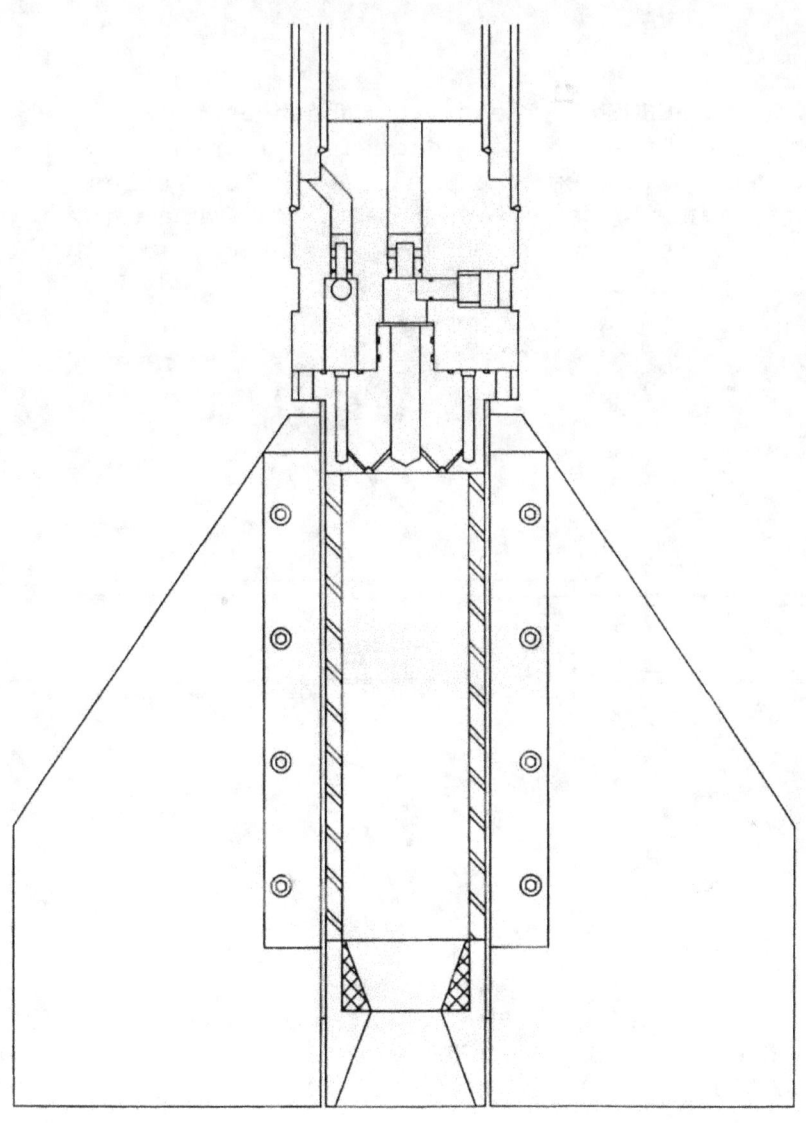

FIN ATTACHMENT: The steel tube housing the ablative liner had four steel angle brackets welded to its exterior, each with four holes sufficient for a 3/8-inch bolt. The aluminum fins were then bolted to these brackets.

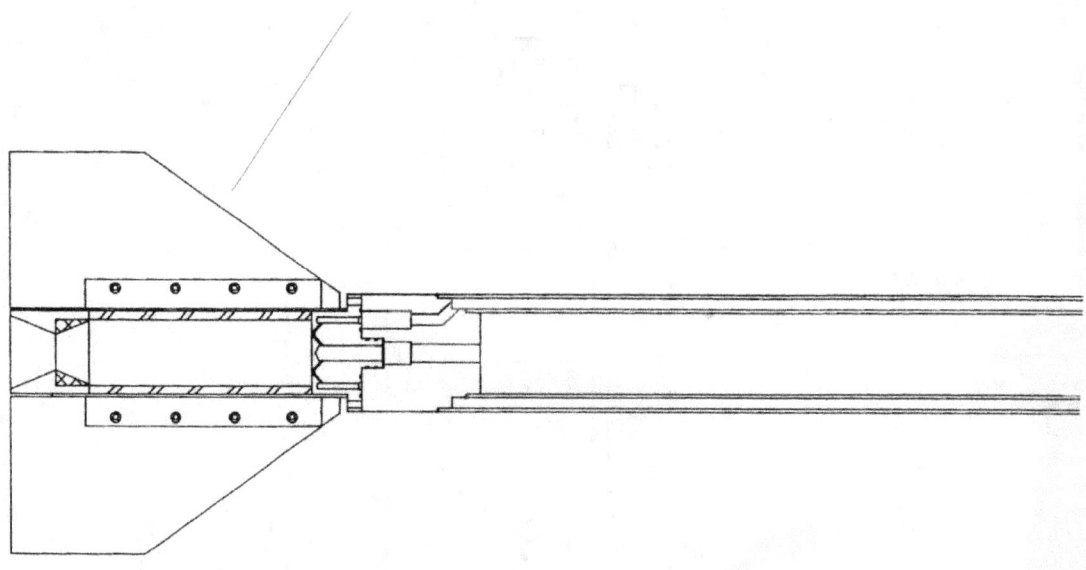

CUTAWAY OF BOTTOM HALF OF ROCKET
SHOWING WELDED BULKHEAD WITH VALVE CHANNELS,
DETACHABLE INJECTOR, ABLATIVE ENGINE, AND FINS

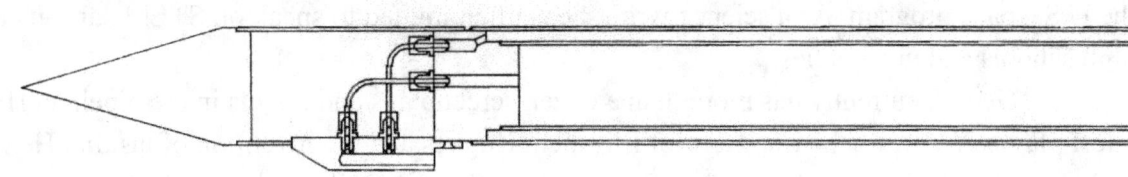

CUTAWAY OF TOP HALF OF ROCKET
SHOWING CONCENTRIC PROPELLANT TANKS,
WELDED FORWARD BULKHEAD, NITROGEN PRESSURE
FEED LINES, CHECK VALVES, PRESSURE FEED CHANNEL,
AND ALUMINUM ADAPTOR TUBE AND NOSE CONE.

George D. Morgan has a BA in Creative Writing from California State University Channel Islands and an MFA in Writing for the Performing Arts through the University of California Riverside's Palm Desert graduate writing program. He is currently finishing a PhD from the University of Texas at Dallas in Visual and Performing Arts.

George has written more than a dozen stage plays and musicals including *Second To Die, Nevada Belle*, and *Thunder in the Valley*. An avid composer, he wrote the songs for the children's musical *The Trial of Goldi Locks*. In 2001 his play *Second To Die* was adapted into a feature film starring Paul Winfield and Erika Eleniak. George's plays and screenplays have won numerous awards, including *Short Line*, which won first place in the International Family Film Festival screenwriting contest.

As the playwright in residence at Caltech, George has written several science-themed plays including *Rocket Girl*, which tells the true story of his mother, Mary Sherman Morgan, and *Pasadena Babalon*, the true story of Jack Parsons, co-founder of the Jet Propulsion Laboratory. His book *Rocket Girl: The Story of Mary Sherman Morgan, America's First Female Rocket Scientist*, has become a best seller. George's follow-up book, *Rocket Age*, covers the history of the U.S. space program. As a science writer he is often invited to speak on STEM subjects to high schools and universities.

George currently has more than a dozen screenplays and novels in the pipeline. His latest play is *Needles*—the true story of Elizabeth Hughes and the invention of insulin. He is an active member of both the Dramatists Guild and the Writers Guild of America. He and his wife Lisa live in Petty, Texas with a cat, a dog and two of their three adopted children.